Date: 4/10/20

Praise for *The Angel and the Assassin*

"*The Angel and the Assassin* is a fascinating deep dive into unsung heroes (and villains) inside our skulls: microglia, immune cells in the brain implicated in everything from depression to Alzheimer's disease to psychosis. This fast-paced investigation presents readers with a behind-the-curtain look at science's quest to unlock the mysteries of these tiny molecules — with huge ramifications for our understanding of brain health and illness. Donna Jackson Nakazawa has a journalist's eye for a story, a scholar's understanding of the research, and a patient's appreciation for how high the stakes truly are."

—SUSANNAH CAHALAN, *New York Times* bestselling author of *Brain on Fire*

"In a stunning show of precision and heart, Jackson Nakazawa offers a captivating, page-turning story of scientific discoveries that overturn centuries of medical dogma and fundamentally reshape psychiatry, medicine, and the treatment of mental and physical illnesses. *The Angel and the Assassin* offers extraordinary promise and heralds hope in a time of skyrocketing rates of 'microglial' diseases — including depression, anxiety, Alzheimer's, and addiction. It is paradigm-shifting reading for us all!"

—CHRISTINA BETHELL, PH.D., M.B.H., M.P.H., professor of Child Health, Johns Hopkins Bloomberg School of Public Health

"*The Angel and the Assassin* is the rarest of books: a combination of page-turning discovery and remarkably readable science journalism. It's a book both to savor and to share."

— MARK HYMAN, M.D., *New York Times* bestselling
author of *Food: What the Heck Should I Eat?*

"I can think of no topic more fascinating or exciting than microglia, the long-misunderstood brain cells whose power over brain health may hold the promise of cure for so many. *The Angel and the Assassin* is riveting, engaging, and visionary."

— TERRY WAHLS, M.D., author of *The Wahls Protocol*

"Engagingly explained and exciting . . . Jackson Nakazawa puts forth a revolutionary new way of thinking about the brain's immune system and its interactions with immune function in the rest of the body. She calls this the 'microglial universal theory of disease,' and the research and emerging treatments based on it have the potential to prevent and relieve depression, anxiety, and Alzheimer's disease, as well as diverse auto-immune disorders. Much of the information here was new to me and has made me more optimistic about the future of medicine."

— ANDREW WEIL, M.D., *New York Times* bestselling author
of *Eight Weeks to Optimum Health* and *Healthy Aging*

"Understanding the capacity of microglia to both heal and harm the brain is one of the most exciting scientific discoveries in recent memory. Donna Jackson Nakazawa brings this heady story to colorful, page-turning, and accessible life. I have great hopes for the practical application of what she reveals."

— AMY MYERS, M.D., *New York Times* bestselling
author of *The Autoimmune Solution*

"Few nonfiction writers can tell the tale of scientific inquiry so vividly that the reader can feel the excitement of discovery with every word. Donna Jackson Nakazawa is one of those writers, and this book tells the tale of one of the most intriguing and groundbreaking discoveries in all of medicine. The tiny brain cell that forms the center of her story will one day affect all of our lives and the lives of our families, and will transform the way mental illness and the ravages of aging are treated."

—SHANNON BROWNLEE, senior vice president, Lown Institute, and author of
Overtreated: Why Too Much Medicine Is Making Us Sicker and Poorer

"*The Angel and the Assassin* introduces us to the smallest cells in our brains—microglia—and compellingly explains startling new discoveries about them that not only change how we think about human suffering—in the form of medical, neurological, and psychiatric disorders from depression and anxiety to Alzheimer's disease—but also have the potential to impact how we understand our minds. Donna Jackson Nakazawa is a powerful storyteller and an expert journalist; her inspiring account of these revolutionary insights will provide a game-changing view of health for generations of researchers, clinicians, and citizens for years to come. As a psychotherapist, psychiatrist, physician, scientist, and educator, I find something in this book to meet each of these role's wildest dreams. Bravo!"

—DAN SIEGEL, M.D., clinical professor, UCLA School of Medicine,
and executive director of the Mindsight Institute

"*The Angel and the Assassin* is one of those astonishing medical yarns that you almost can't believe: how the power of this tiny cell was so long overlooked, how integral it has become to our understanding of neuroscience and immunology, how it has transformed the most basic ideas of who we are as humans. The book is especially essential reading for women, who face depression, Alzheimer's disease, and autoimmune disorders at higher rates than men."

—PEGGY ORENSTEIN, author of *Girls & Sex: Navigating the Complicated New Landscape*

"*The Angel and the Assassin* is a captivating look at the new science altering our understanding of a range of conditions that affect women especially—from depression and anxiety to Alzheimer's disease and chronic pain. Upending centuries of mind-body dualism in Western medicine, this book is a powerful reminder that so much of human physiology is still a mystery."

—MAYA DUSENBERY, author of *Doing Harm: The Truth About How Bad Medicine and Lazy Science Leave Women Dismissed, Misdiagnosed, and Sick*

"*The Angel and the Assassin* is a must-read for everyone interested in the new science of brain health. Get ready to make friends with your microglia."

—LOUANN BRIZENDINE, M.D., author of *The Female Brain* and *The Male Brain*

"We are in a new era of brain health, one that holds new possibilities for protecting and repairing the brain. For the field of psychiatry in particular, our ability to identify and treat neuroinflammation is set to be a real game-changer. Donna Jackson Nakazawa eloquently captures these impactful advances. *The Angel and the Assassin* is an impressive, inspiring, and a timely call to arms."

—SUSANNAH TYE, PH.D., director of Mayo Clinic's translational neuroscience laboratory and senior research fellow, Queensland Brain Institute

"*The Angel and the Assassin* is a provocative and eloquent account of cutting-edge research that proves the intimate connection between mind and body and that has the capacity to transform our thinking about traumatic stress syndromes, addiction, depression, and dementia. The book not only provides hope and a way forward for many who have suffered greatly, but also sets the stage for a paradigm shift for clinicians and researchers in the decades to come."

—RUTH LANIUS, M.D., PH.D, professor of psychiatry, Harris-Woodman Chair in Mind-Body Medicine, and director of the Posttraumatic Stress Disorder Research Unit at the University of Western Ontario, and coeditor of *The Impact of Early Life Trauma on Health and Disease*

"Donna Jackson Nakazawa has written a deft scientific story about microglia, the 'Cinderella' cells of the brain. It turns out that microglia, once thought to be simple housekeepers, may have a starring role in a range of disorders from depression to dementia. Revealing the breakthroughs that give us a new view of the brain as an immune-system organ, Jackson Nakazawa also explains the possible translations of the science into solutions for brain disorders, health, and disease."

—THOMAS INSEL, M.D., president of Mindstrong Health
and former director, National Institute of Mental Health

The Angel and the Assassin

THE
Angel
AND THE
Assassin

THE TINY
BRAIN CELL
THAT CHANGED
THE COURSE OF
MEDICINE

DONNA JACKSON
NAKAZAWA

BALLANTINE BOOKS

NEW YORK

Published in the United States by Ballantine Books, an imprint of
Random House, a division of Penguin Random House LLC, New York.

BALLANTINE and the HOUSE colophon are registered trademarks of
Penguin Random House LLC.

LIBRARY OF CONGRESS CATALOGING-IN-PUBLICATION DATA
Names: Nakazawa, Donna Jackson, author.
Title: The angel and the assassin : the tiny brain cell that changed the
course of medicine / Donna Jackson Nakazawa.
Description: New York : Ballantine Books, [2020] | Includes
bibliographical references and index.
Identifiers: LCCN 2019034927 (print) | LCCN 2019034928 (ebook) |
ISBN 9781524799175 (hardcover) | ISBN 9781524799182 (ebook)
Subjects: LCSH: Microglia.
Classification: LCC QP363.2 .N35 2020 (print) |
LCC QP363.2 (ebook) | DDC 612.8/1046—dc23
LC record available at https://lccn.loc.gov/2019034927
LC ebook record available at https://lccn.loc.gov/2019034928

Printed in Canada on acid-free paper

randomhousebooks.com

2 4 6 8 9 7 5 3 1

FIRST EDITION

Book design by Simon M. Sullivan

For Claire

Contents

The Angel and the Assassin

When the Body Attacks the Brain

M Y INTEREST IN THE MYSTERY SURROUNDING THE LINK BETWEEN physical disease, the immune system, and brain-based disorders began more than a decade ago, when I developed a rare autoimmune disease that left me unable to walk. Over the course of a five-year span, between 2001 and 2006, I spent, all told, a year in bed or in a wheelchair.* My doctor, a neuroscientist at Johns Hopkins, was studying my disease, Guillain-Barré syndrome. My case was unusual in that I had been felled and paralyzed not once, but twice by the disease — rare but not entirely unheard of.

My doctor, who knew I was a science journalist, and I talked at length about what was happening repeatedly to my body, and how he hoped to reverse my paralysis. He explained that in my disorder, as in all autoimmune diseases, my immune system's white blood cells were behaving erratically, like an army gone rogue. Instead of judiciously protecting my body from invading pathogens, my white blood cells were mistakenly attacking and destroying the protective myelin sheaths that coated my

---

* Readers of my previous work may be familiar with my story; if so, forgive any brief repetition here.

body's nerves, causing the nerve-muscle interconnections that I needed to stand and walk, or simply wriggle my feet, to go dark.

I began to think of these overactive immune cells as being like Pac-Men—those 1980s video game characters—crazily gobbling up and destroying my good nerve cells, eating away at the crucial nerve connections that made my physical body mine—strong, capable, dependable me.

Regular intravenous therapies would, my neurologist hoped, reboot my immune system so that my white blood cells would stop overreacting and behave normally. If this approach worked, and we could get my overvigilant immune cells to back off from their attack, my nerves would hopefully start to regenerate on their own. The nerves that had gone dark would light up again, regrowing enough neuromuscular connections that I might even be able to walk once more.

Many months later, it would turn out that my neurologist was right. In the end, my immune system's overexcited cells did back off. I did get much better. Not all of my nerves regenerated, but enough did so that I could walk again, and the significant gains I made have allowed me to live a good life. The human body can be miraculous that way.

I had one question for my medical team during that time, however, that they could not answer. After losing the use of my legs, I'd also experienced what seemed to be some distinct and disquieting cognitive changes. For one thing, although I'd always been pretty even-keeled, I found myself facing a black-dog depression. The feeling was at times so oppressive that when I read *Harry Potter* aloud to my young children, I felt as if I'd been attacked by the "dementors"—those dark, sky-drifting ghouls who introduce a cloud of despair that steals a human being's happy thoughts and replaces them with bad ones. It almost seemed, I told my primary care physician, "like someone has inhabited my brain."

I'd also always had a stellar memory and could recall conversations that had taken place weeks, months, even years earlier, word for word. Now, though, I was having to write simple things down, such as what time the physical therapist was coming the next day. *No big deal,* I thought. *That's new for me, but it happens to lots of people.*

But there were other, more worrisome "glitches." It was, for instance, taking me a disturbingly long time to recall the names of people I'd lived with and loved all my life. At Thanksgiving dinner in 2005—when I'd graduated from being bedridden to getting around, albeit in a limited fashion, with a walker or cane—it became a source of good-natured amusement for my extended family to watch me struggle to come up with a person's name. Everyone smiled patiently as I called out a string of names across our long table, one after the other, until I landed on the right one: "Sam! Christian! Zen! Don! Jay! Cody!"—the dog's name— "Chip! Could you please pass the salt?"

Even that could be seen, I told myself, as almost humorous; at least my brain was able to distinguish between male and female names. But some things weren't at all funny. My six-year-old daughter would ask for help with simple first-grade math and I found my brain stuttering just to add seven and eight. Or I'd reach down to tie her shoes, something I'd done for years, and find myself staring dumbly at the laces, struggling to re-member how, exactly, it was done. I can still recall cutting up slices of watermelon, putting them in a bowl, and staring down, thinking, *What is this again?* I *knew*, but I could not remember the word. I'd cover my lapse by bringing the bowl to the table and waiting for my children to call out, "Yay! Watermelon!" And I'd think, *Yes, watermelon, of course it is.*

My rife anxiety was less surprising, given that I'd been plunged, practi-cally overnight, into almost complete physical paralysis and hospitalized for several weeks at a time, more than once. In the process, as my body's nerves had demyelinated, I'd endured excruciating muscle spasms, and undergone repeated spinal tap and electrical conduction tests in which technicians shocked nerves in my arms and legs to see which ones were not responding. For a time, I'd lost use of the muscles needed to swallow solids. In the hospital, I'd blacked out after an allergic reaction to an IV therapy and come back to consciousness to find a team of ashen-faced doctors hovering over me with a crash cart, and a nurse uttering prayers under her breath. And then there were the endless weeks I'd spent in physical rehabilitation centers, learning to balance myself and push a walker across a room on numb legs that were nevertheless riddled with shooting pains.

But this omnipresent anxiety felt like a separate disease in and of itself.

My memory, clarity of mind, word recall were different—my brain did not feel like my own.

I sometimes commiserated on the phone with a friend of mine, Lila, who had Crohn's disease and was having similar cognitive concerns. She told me that one day when she'd dropped off her older son at kindergarten, she'd left her two-year-old son (he'd clambered down out of her arms to play at the sand table while she chatted with the teacher) in the classroom too, by mistake. The teacher had followed her out of the class, with the toddler shrieking in her arms. "I forgot my own son!" Lila said, sobbing. Her doctor had sent her to a psychiatrist, who was prescribing antidepressants for anxiety and obsessive-compulsive disorder and Ritalin for ADD. But, she wept, "I never needed Zoloft or Ritalin before I had Crohn's."

I empathized with Lila; her concerns mirrored my own. The next time I saw my primary care physician, I confided to her, "It's like a part of my brain has gone dark too, along with my legs and feet." It reminded me of people I'd known who'd had a mini stroke. "What's going on with that?" I asked, hoping to learn more about my own situation.

I had not had a mini stroke, my internist assured me. She reminded me that my life had been much changed, and what had happened to me had been traumatic. It made sense that my mental state would be deeply affected. My neurologist, meanwhile, encouraged me, cheered me on in my recovery, and reassured me that as time went by I would continue to heal.

And so I did. Still, a number of my cognitive symptoms lingered. And I could not shake the feeling that just as my body had been altered, something physical had also shifted in my brain.

I began to wonder whether any immunologists, who studied the immune system's effect on all the other systems and organs of the body, suspected there might be biological links between physical immune dysfunction and brain-related or psychiatric illness. I began to dig a little deeper.

During this same time, between 2007 and 2010, I was also enormously busy writing books, lecturing, managing my illness, and raising a young family. But I kept doing sideline research on the topic. And it turned out that a handful of researchers in labs around the world were setting forth to examine this precise link. I collected every peer-reviewed study that documented patients whose immune systems were overactive — thus causing inflammation and disease in their body—who also reported considerable cognitive and mood issues.

In 2008, research revealed that patients with MS also experienced changes in their ability to remember things and were several times more likely to suffer from depression and bipolar disorder than individuals without MS.

In 2010, an analysis of seventeen different studies showed that having lupus, which often manifests in systemic inflammation in the organs of the body, was associated with a much greater likelihood of suffering from depression, and even psychosis. Shockingly, as many as 56 percent of patients with lupus reported cognitive or psychiatric symptoms, including trouble concentrating, mood disorders, depression, generalized anxiety, and learning issues. Having lupus is also associated with early dementia.

And, that same year, researchers found, after looking at thirty years of health data on three million people, that individuals who'd recently been hospitalized with bacterial infections were 62 percent more likely to develop depression, bipolar disorder, and memory issues.

Several case studies in the scientific literature showed a link between disorders in the bone marrow—where most of our body's immune cells are "born"—and schizophrenia. In one case study, a patient who received a bone marrow transplant from his brother, who suffered from schizophrenia, also developed schizophrenia just a few weeks after receiving the bone marrow. In another case study, a young man with schizophrenia and acute myeloid leukemia received a bone marrow transplant from a healthy donor, and both his cancer and his schizophrenia were cured.

Yet, as compelling as all this research was, at the time that many of these studies were coming to light, it just did not make any scientific sense that being sick in the body could be connected to, much less cause, physical sickness in the brain.

. . .

Let's take a minute and step back to remember high school biology basics about how your immune system functions. Your white blood cells—the army of the immune system—constantly circulate in your body, scanning for invaders, germs, pathogens, and environmental toxins you can't see or smell. In the time it will take you to read just this sentence, your immune system will respond to thousands of unseen threats to your well-being. The germ cloud from the person sitting next to you on the bus who just sneezed. The minute bacteria in the dirt still clinging to the organic salad you had for lunch. The fungus that's growing in your office building's HVAC system. The chemicals coming off the new plastic bins you bought to file away this year's tax receipts. Your white blood cells are there to snuff out an endless parade of invaders, 24-7.

If you cut your thumb while slicing onions, your immune system's white blood cells rush in like a SWAT team to battle any infiltrating bacteria while also mending damaged tissue. Your thumb swells up and becomes inflamed—red, hot, swollen, and painful—while your white blood cells congregate to do the necessary repair work. While red, hot, swollen, and painful might seem problematic, this is actually your immune system doing its job, and doing it well.

But inflammation occurs in ways that can be harmful too. If your body happens to be overtaxed by too many environmental triggers, your army of white blood cells can become overwhelmed and go into overdrive, mistakenly attacking your own tissues, joints, organs, and nerves, leading to autoimmune diseases such as rheumatoid arthritis, lupus, MS, and type 1 diabetes.

Inflammation and autoimmune disease can both arise in virtually any organ or system in the body. White blood cells have to get it just right. If they don't battle hard enough, infection and pathogens can spread and shut down organs, and you can die of sepsis. If white blood cells overdo it, they may protect you from outside invaders, but in the process mistakenly launch an inflammatory hyperattack on your body too, giving you a new illness you didn't previously have. (In my case, when I fell ill, I'd originally had a stomach virus. My white blood cells knocked out the infection—but went too far and knocked out the myelin sheaths around my nerves too, which resulted in Guillain-Barré syndrome.)

There is only one organ in the human body that scientists have categorically believed for over a century was *not* affected by the body's immune system.

That organ is your brain.

And, it follows, if your immune system can't reach—and doesn't rule—your brain, then your brain cannot be touched by inflammation or become inflamed.

When my internist and I talked about the cognitive changes that occurred during flare-ups of my physical disease, we assumed these symptoms had to be purely emotional in nature because for decades, textbooks had declared that when a disease was attacking the human body, the brain was off limits. The prevailing idea in neuroscience, long taught in medical schools, was that the brain was "immune-privileged." Scientists unilaterally believed that, for better or for worse, the immune system had access to every single organ in your body except your brain. Inflammation in the brain could occur only if there was an external event—head trauma, or an infection known to directly target brain tissue, such as meningitis. Other than that, the brain simply could not generate inflammation.

This theory made sense for anatomical reasons. After all, if your inflamed thumb swells up—even to twice its normal size—your skin can expand (albeit painfully) to accommodate the internal swelling. But if your brain expands, it has nowhere to go—it's trapped inside your skull. If too much pressure builds up, the brain—and you—can't survive. In extreme instances, when there is a major head trauma such as in a car crash, brain tissue can swell—and pressure has to be relieved by surgeons drilling into the skull. So early anatomists had very good reason for assuming the brain was simply not an immune organ.*

---

* This long-standing belief that the brain was immune-privileged also had to do with scientists' understanding of the blood-brain barrier, a dense, complex constellation of cells that congregate around the blood vessels that lead to the brain. These vessels are so tightly packed together that they block particles from the bloodstream and the body,

But by 2011, an increasing number of researchers were starting to doubt that dogma. Neuroscientists and immunologists were beginning to wonder: Could the brain be affected by inflammatory processes—and if so, how?

But these questions had no answers.

Occasionally, I'd mention what I called my Brain Flame project to my literary agent. She'd play devil's advocate, asking, "But if we don't know *how* or *why* the brain could be affected by the body's physical immune system, how can you say that having an overactive immune system can result in brain-based illness?"

To be honest, at that time—around 2011 and 2012, prior to the groundbreaking discoveries I'll detail in the following pages—I could not prove that the brain was affected by disorders of the body. No one understood the exact way in which immune dysfunction in the body could lead to brain-related psychiatric or neurodegenerative disorders, or cognitive decline. The brain was immune-privileged, and that was that.

And yet, even as my agent and I had that conversation, the scientific community was on the verge of exploding with surprising answers to the questions that had haunted me for more than five years. Newly understood scientific links between physical immune health and brain health were emerging at a rapid pace.

And all of these discoveries revolved around one type of tiny cell that had, for over a century, been largely overlooked in human health. These deceptively minuscule cells, called *microglia*, had never been seen as significant in determining mental and cognitive health. Then, in 2012, groundbreaking research showed that, contrary to scientific dogma, these microglial cells held enormous power to protect, repair, and repopulate

---

including immune cells, from reaching the brain. The inviolate nature of the blood-brain barrier has long been seen as proof that the brain is off limits to the body's immune system—and therefore immune-privileged.

the brain's billions of neurons and trillions of synapses, or to cripple and destroy them, leaving wildfire-like devastation in their wake.* Unbeknownst to researchers, microglial cells had long been *functioning as the white blood cells of the brain*, governing the brain's health.

In fact, the five-year span between 2012 and 2017 was a watershed time for neuroscience and immunology, during which the discovery of microglia's true role in brain health would cause the two fields to become one.

During these same years, scientists also unveiled that these microglial cells in the brain were chatting with the body's immune cells in direct and indirect ways: If there was inflammation in the body, there would almost inevitably be immune changes within the brain. Moreover, these immune changes in the brain could manifest themselves as cognitive or neuropsychiatric disorders.

These immune changes could affect the brain's synapses and neural connections even when there were no signs of physical illness in the body itself.

And this meant that we could begin to understand the interior health of the brain in ways that had been unimaginable just a few years earlier. The discovery of microglia's ability to both build and break down the brain provided researchers with an entirely new organizing principle for deciphering brain health and brain illnesses.

Still, it was clear that even as this research rocked the scientific world, patients were not learning about what scientists now knew, or why this information was so vital to their well-being.

As a science journalist, I had often seen that it took many years for research to trickle down to the patients who needed answers most. My goal as a writer was simple: report on findings that could ease suffering, in order to close the gap between what science knew and what patients needed to know in order to live their best, healthiest lives.

And so I set out to do a journalistic investigation, delving into, analyz-

---

* Synapses are the small gaps between neurons that allow electrical and chemical signals to pass from one neuron to the next.

ing, synthesizing, and connecting the dots between the stunning new stories of research emerging on microglia and their role in helping to cause, but also to reverse, brain-based illness.*

This book tells the story of the birth, exploration, and promise of one of the most paradigm-shifting and powerful stories in the history of medicine—the story of these tiny microglial cells in brain health. In the pages that follow, we'll see how these cells hold the potential to transform human health by helping us to repair the brain in ways that we have previously only dreamed of. I'll take you on a journey of discovery, detailing the most exciting promise for healing that I've ever reported on in my thirty-some-year career as a science journalist.

I believe that you will come to see, as I have, how and why this groundbreaking revelation about microglia has forever overturned many of the long-standing and powerful assumptions we have made about the mind-body-brain connection—including why it tells us, categorically, that the brain is indeed an immune organ, ruled by these powerful, enigmatic immune cells. Along the way, we'll follow the progress of a few patients whose lives have been transformed by our new understanding that this cell serves as both the angel and the assassin of the mind.

Perhaps most important, we'll learn about cutting-edge new approaches that help reboot and redirect the activity of microglia so that these little immune cells stop their attack, thus allowing the brain's neurons and synapses to regrow and heal.

---

* It may appear that I'm not paying sufficient attention in these pages to how our thought patterns, emotions, feelings, and habits of rumination can affect our brain circuitry over time, and how unlearning these negative patterns through practices such as cognitive behavioral therapy and mindfulness training can change the brain, leading to positive synaptic changes. Readers of my work will be aware that I've devoted two previous books, *Childhood Disrupted* and *The Last Best Cure*, to the latest research in the field of psychoneuroimmunology, dealing with how our life experiences, past traumas, patterns of thought, and habits of mind profoundly affect our brains, our immune systems, and our present lived emotional experiences.

We'll peer too inside top research labs across the country and see how the work of a small band of fearless and dedicated neurobiologists, who discovered the extraordinary power of microglia, has changed the future of human health forever.

Including, quite possibly, yours.

ONE

The Accidental Neurobiologist

As you enter Beth Stevens's lab office in Boston, Massachusetts, where she serves as associate professor of neurology at Boston Children's Hospital and Harvard, you're greeted by a giant whiteboard. At its center sits an elaborate hand drawing of a microglial cell, rendered in bright green fluorescent marker. Tentacle-like arms extend probingly out from the cell's blob-like center, each delicate arm pointing toward a different handwritten list of the primary research projects currently underway in Stevens's lab, along with important deadlines. It's clever.

It's nearly five o'clock in the afternoon. Stevens's ten-year-old daughter, Riley, is finishing homework at her own child-sized desk not far from her mom's. Riley's hair—the same towheaded white-blond as her mother's—hangs in neat pigtails. Riley pushes her glasses up on her nose, heads over to the majestic whiteboard, and picks up a marker and dry-eraser. Her blue eyes—also her mom's—sparkle mischievously.

"Riley, don't erase that board!" Beth calls out. Her voice conveys pretend mom sternness. "All kinds of crazy things are going to start happening around here if that board gets erased!"

Stevens's husband, Rob, walks in just then to pick Riley up after school. He offers me a warm hello, then smiles playfully as he points out a silver espresso cup sitting on Stevens's desk. The cup is still steaming.

"Yup, I just had another cup of espresso," she says, exchanging a smile with her husband. A large mug with the words DEATH WISH COFFEE also sits on her desk. She looks at me and gives a minuscule shrug. "These mugs are gifts from my lab. I guess that should tell you something?"

She kisses her daughter and husband goodbye before showing me around the lab. Stevens is stylish and crisp in an olive-green summer dress, her wavy blond hair neatly pinned back with a silver clip. She gestures under her desk, where research papers rise in foot-high stacks. "My reading pile!" she laughs. Above her desk, photos of her daughters, Riley and Zoe, are pinned to a bulletin board, interspersed with favorite pieces of their preschool artwork. There's a photo of the beach house where she and Rob vacation every summer on Cape Cod. Stevens points fondly to a photo of herself embracing a young woman in a graduation cap and gown. They're both smiling widely. "This is my first grad student." There are several collages made up of the faces of dozens of the students and colleagues she's worked with over the past twenty years. "Looking at all these faces makes me really happy when I'm feeling stressed," she says.

In the area outside her office sits an espresso machine, which Stevens gifted to her lab. She points out the "dual heads so two people can fill up at the same time." (A fellow neuroscientist once described Beth Stevens as being a lot like "a four-shot espresso." It's an apt description.)

A plate of cookies awaits kids who—like Riley—might come to the lab for a few hours after school to do homework while they wait for their mothers. (Yes, much of Stevens's team is female.)

Intermingled with the microscopes and computer screens there is one apparatus I've never seen before in a biology lab: a miniature brewery. "My postdocs and students brew our own beer," Stevens explains, with a laugh. "We call it microgliale."

It's a busy, cozy, happening, caffeine-fueled, fun place, Stevens's lab. Today, there are fifteen postdocs and students working on different projects. Stevens runs a smaller second research group at the Broad Institute, a biomedical and genomic research center, and she is in high demand at neuroscience conferences all over the world to share her game-changing discoveries about the tiny cells that science almost forgot—microglia.

But it didn't start out that way.

In many ways, Beth Stevens is an accidental neurobiologist.

Student of Nature

Beth Stevens grew up in the small industrial city of Brockton, Massachusetts, known for its history of shoe manufacturing. Her father was an elementary school principal in downtown Brockton, and her mother taught at another local elementary school, closer to their home. Reading books and kitchen table arithmetic were both encouraged and supported. Stevens was bookish, like her family, but she also had a more hands-on brand of curiosity.

She spent hours in her backyard turning over rocks, sitting in trees, looking at the undersides of leaves, smearing sap between her fingers, and watching insects, in an effort to discern the unseen workings of the natural world.

Later, in middle school, when it came time to participate in the frog dissection that most students dreaded in biology, Stevens felt none of the squirmy hesitation of her classmates. "I couldn't imagine anything more intriguing than seeing how the inside of a frog's body worked," she says, taking a sip of her espresso as we sit at her desk. After that day, "I know it might sound gross, but if I saw a dead squirrel or opossum on the side of the road—yes, roadkill, it's awful!—I'd poke gently at it with a stick, just trying to peer inside. I wanted to understand how its body functioned, and why it died."

To young Beth, it seemed as if looking inside things was the most important and interesting thing you could do in the world.

But there were no scientists in her family. When she did read about a biologist having discovered something exciting, it was invariably a man. She had the sense, growing up in her town, that she was a bit odd—one of those things that is not like the others, as the saying goes. "It certainly never occurred to me that my interests could lead to a career," she recalls, looking back.

That began to change when Stevens took an Advanced Placement biology course in high school. Her teacher, sensing her interest, told her stories about past female students who had gone on to become researchers. He held down a second job working in a medical lab, and he sometimes brought projects into class. "We'd be pouring different mediums into petri dishes and turning on Bunsen burners and I'd think, *Wow, can you really do this as a job?*" Beth says.

When she graduated from high school in 1988, Stevens went on to study biology and medical laboratory science at Northeastern University in Boston, sure she'd later go to medical school. One term, she interned in a hospital lab, where she assisted researchers in identifying a food poisoning outbreak: *Listeria monocytogenes* bacteria lurking in store-bought sausages.

After Stevens graduated, she wanted to find work that would help build her résumé while allowing her time to study for the MCATs. Her husband, Rob Graham, then her boyfriend, had landed a job working on Capitol Hill in Washington, D.C. Stevens was seeking lab experience — and one of the best and biggest labs in the world was situated on D.C.'s outskirts: the National Institutes of Health.

It was 1993. "We moved to D.C. and I thought I'd wait tables at a Chili's restaurant near NIH for a few months until I landed something," Beth recalls. "On breaks I'd pull off my apron and run over to NIH to search the job board and drop off my résumé." Stevens liked to read science journals in her spare time, and she'd recently read "the very odd and fascinating case of a woman with a parasitic infection inside her eye," she tells me. "So I thought I'd really like to work in infectious disease." Among the dozens of applications she put in, she submitted one to work for a Nobel laureate who was studying infectious disease and HIV.

Ten months into her job search, Stevens got a call from the HIV lab offering her the position of lab tech. She was twenty-two years old and had, she thought, landed her dream job. But "I got another call around that time too—from a scientist whose lab I hadn't applied to," she says, with some bemusement.

Doug Fields was, at that time, a young neurobiologist who was setting up his first lab at NIH. He'd called Stevens out of the blue. "He told me he'd thumbed through the rejected résumé pile in NIH's personnel office, where mine had landed." He explained that he was studying the firing patterns of neurons, and how that affected brain development.*

"Going into neuroscience simply was not on my radar at the time," says Stevens. Besides, to someone fascinated with viruses and infectious diseases, it seemed less interesting than studying HIV. So Stevens turned Doug Fields's offer down.

Then life took a circuitous turn. "I showed up for my first day of work at the Nobel laureate's HIV lab, and the lab manager told me there was a hiring freeze; they'd forgotten to tell me I no longer had the job," Beth says. "I went home feeling more dejected than I'd ever felt in my life. The next day I put my waitress's apron back on and went back to serving burritos at Chili's. After almost a year of looking, I'd only had two job offers." She laughs. "And I'd turned one of them down!"

Rarely one to give up, Beth Stevens devised a wacky plan. "I had too much pride to call Doug and ask if his job was still open; I knew I'd feel so embarrassed if he'd filled it." So Beth and Rob sat down at the kitchen table in their rented D.C. studio apartment, and called Doug's lab. Rob pretended to be a recent college grad with a biology degree looking for his first job. Doug asked Rob a few technical questions about his (made-up) lab experience. Rob kept writing notes and pushing them across the table to Beth—"What does *this* mean?" Beth scribbled answers and pushed them back. When Rob hung up the phone he told Beth, triumphantly, "He's still looking for a lab tech!"

Stevens laughs, recalling this small moment of subterfuge.

"I waited two whole weeks, so that I wouldn't seem so transparent, and then I called Doug and asked him if he was still hiring." And miraculously, Beth says, "He offered me the job for a second time. Taking it was one of the best decisions of my life."

---

* At the time, Fields was studying how neuronal firing regulated neuronal gene expression and development.

Lab Fever

Stevens got right to work in Doug Fields's lab. Fields asked her to focus on growing myelin—a fatty white substance that surrounds and protects neurons. "I spent weeks reading papers and calling scientists who knew how to grow myelin to get advice," she says. When Stevens finally got myelin to grow in a petri dish, "It was so gorgeous I couldn't stop staring at it. At that moment, I got hooked; I caught the science bug."

Meanwhile, Fields was beginning to turn his attention to identifying the function of a type of cell involved in brain development, known as Schwann cells. Schwann cells were part of a group of little-studied non-neuronal cells in the brain, known as *glial* cells. Glial cells, scientists universally agreed at that time, played a rather unglamorous role in the brain, simply supporting neurons and synapses.

Fields was interested in Schwann cells because in the developing embryo, they helped myelin to grow, which helped neurons. Neurons were the undisputed flashy darlings of the scientific world, and by 1993 had been for more than a decade. Neurons were responsible for creating the trillions of synaptic connections needed for individuals to think, feel, remember, learn, and love, for all of our intuitions and perceptions.

Neurons were seen as the A-team players of our mood, our mental health, our memory. Glial cells made up the B-team; they catered to the needs of neurons the way an entourage caters to the whims of a movie star.

Every embryo starts out as a ball of stem cells. As the fetus develops in the womb, stem cells begin to "differentiate" into different types of cells, creating the different organs and systems of the human body. For instance, some stem cells become keratin cells, forming hair and nails. Other stem cells become organ cells, developing into heart or liver tissue. Others become nerve cells in the body—or neurons in the brain.

The small family of non-neuronal brain cells known as glial cells included four types of cells, including Schwann cells. They had varied origins during development, and researchers were still investigating their different roles in the brain. Glial cells called *oligodendrocytes* were known to originate, like Schwann cells, in the developing human embryo from the same family of stem cells as nerve cells; they also supported the growth of myelin. A third type of glial cell, *astrocytes*, originated from the same family of stem cells as neurons; they influenced the growth of neurons and synapses.

But there was a fourth type of glial cell that scientists had little interest in at all in terms of normal brain function: *microglia*.

"People just didn't think microglial cells were terribly important in terms of a healthy, developing brain at that time," Beth explains, looking back. In fact, one of Beth's frustrations in getting myelin to grow in the lab, and in other petri dish experiments, had to do with the fact that her petri dishes would often become contaminated by microglial cells. "Microglia *ruined* my experiments," Stevens says, smiling at the irony of it all now. "They'd get in my cultures or on my slide when I was trying to look at other cells. And I'd be groaning to myself, 'Oh no no no, here are these annoying little microglia *again*.'"

Stevens wasn't alone in her disdain for this tiny cell. At a time when funding for brain cell research was really ramping up at NIH, microglia were the one type of brain cell that didn't seem to be on neuroscientists' radar, period. When scientists did talk about microglia, they were, like Stevens, usually cursing them.

And what scientists knew about them was basically twofold: They were tiny (hence the "micro" in their name) and they were boring.* Microglia appeared to have one job: If a neuron died, they carted it away. They also disposed of cells that naturally die off as we age. They were the brain's humble trash men. Robot-like housekeepers. End of story.

---

* Microglia were first named and described in the early 1900s by Pío del Río Hortega, a student of Santiago Ramón y Cajal, widely considered to have been the world's first neuroscientist. When Hortega first noted these cells, he called them *microglia*, grouping them with the three other types of glia. Later, in the twentieth century, scientists looked more closely at microglia and their role in brain injury and infections that directly targeted brain tissue, such as meningitis. Margaret M. McCarthy, Ph.D., researcher and professor of neuroscience at the University of Maryland School of Medicine, puts it this way: "Microglia were only considered in the context of injury, or doing normal housekeeper jobs, and were thought to be otherwise 'quiescent,' meaning they sat inert and did nothing unless an injury occurred." No one was thinking about microglia in terms of how a healthy brain functioned, or in terms of normal human brain development.

Still, Stevens was struck by the fact that these annoying little house-keeper cells also made up a really big part of the brain's population—more than one-tenth of our brain cells. Despite their numbers, few scientists had spent any time taking note of their activity, so no one was observing what this minuscule cell was really up to. "It was such a mysterious part of neuroscience," she says.

Stevens made a mental note, then tucked her curiosity about microglia away.

Meanwhile, she was still planning to go on to medical school. One day when Fields saw her studying for the MCAT, he told her that when she got into med school, he'd need to find someone to take her place. "In that moment I realized I didn't want anyone to take my job," she says. "I was more excited by the idea of examining life under a microscope in the lab than examining patients in the clinic."

Moreover, Stevens thought the work they were doing had the potential to someday lead to "a new understanding of disease pathways," she says, very possibly in ways that could help "prevent or treat disease in patients." Stevens withdrew her medical school applications and became Fields's lab manager. With Fields's support, in 1994, she began working toward her Ph.D. in neuroscience at the nearby University of Maryland.

Some nights, between running experiments, managing the lab, and writing her Ph.D. thesis, Stevens would be so stretched for time, and so concerned about watching over her projects in the lab, she would make "a little bed out of extra clothes and coats and sleep on the floor under the conference table, so that I could keep checking to see if there was any progress in my experiments."

Nearly ten years after joining Fields's lab, in 2003, Beth Stevens graduated with her doctorate in neuroscience.

The Housekeeper in Disguise

In 2004, Stevens was invited to pursue her postdoctoral studies at the lab of another leading glial biologist, Ben Barres, Ph.D., at Stanford. Barres was arguably the leading glial researcher in the world. At the time, he was looking primarily at astrocytes, and how they influenced neurons. But he too had recently become intrigued by the question of what microglia were really up to in the brain. His interests aligned well with Beth's. Besides, it was professionally time to make a leap—staying at the same place for your postdoc is considered, as one scientist put it, "career suicide"—so Beth and Rob, now married, packed up and drove across the country to Palo Alto, California.

Right around that time, researchers had developed a powerful new method to peer into the brain and take high-resolution moving images—enlarging images of even the tiniest cells to gigantic visual projections. A fellow researcher visiting Barres's lab that year gave a talk during which he showed a beautiful set of imaging studies that offered a far better view of cells within the brain than anything Beth Stevens had ever seen.*

"These imaging studies showed microglia *in action* for the very first time," Beth recalls. "Suddenly, we could *see* microglia in the brain; we had this new visual tool that allowed us to watch them." Stevens gets up from behind her desk, where we are talking, and scoots her chair closer to a computer terminal to pull up very early footage of microglia. She leans toward the computer screen, pointing with the eraser end of her pencil to the swirling microglia dancing in the brain. What I'm looking at reminds me of images I've seen of the Milky Way—if all the stars were glowing green and whirling in large, purposeful clusters swimming against a black sky.

"The first time I saw this I sat there stunned," Beth says. "I could see

---

* Axel Nimmerjahn, Ph.D., who presented this work, was also mentored by Ben Barres. Nimmerjahn was one of the first scientists to image microglia in the living mouse brain; this was, says Stevens, "a game changer" in the study of microglia.

all these bright green microglial cells that almost appeared to be *moving* through the brain," she says.* "They were highly active. And when there was any damage to the brain, like a brain injury, the microglia would send their long armlike protrusions rushing toward the site. I kept thinking, *Whoa, what are these little cells* doing? *They're so dynamic! They're everywhere! No other cell in the brain can do this. How could we have ignored these cells for so long?*"

A Common Origin

Meanwhile, other scientists were getting more curious about microglia too.

Researchers at Mount Sinai School of Medicine in New York were asking: When do microglia first appear in the developing embryo? Very early on, it turned out. Microglia emerged from the same family of stem cells that went on to develop into the immune system's army of white blood cells and lymph cells. But instead of staying in the body like white blood cells, on the ninth day of gestation, microglia travel up through the bloodstream and into the brain—*where they settle and stay throughout a person's entire lifetime.*

In other words, microglia and white blood cells have the same origin story. They are close cousins in the immune system's police force, yet microglia reside in an organ that had long been thought to be completely untouchable by the body's immune system.† It was true that white blood

---

* Although in these images it appeared to the naked eye that the microglia themselves were racing and swirling around the brain, it is really more the case, in strict scientific terms, that it is what scientists refer to as microglial "processes" that were moving. What this means, in simple terms, is that microglial cells themselves don't move much. Rather, microglia tile themselves throughout the brain so that they can monitor different tiny sections of the brain. When microglia reach out their long arms to check on neurons, they are checking on neurons in only that specific area, and their fine, elongated branches extend and retract and move very quickly. This is similar to the way in which neurons relay signals across synapses, but the neurons themselves don't move.

† According to Margaret McCarthy, when Río Hortega first named microglia, the term was an unfortunate misnomer, given that microglia are indeed immune cells,

cells didn't have access to the brain—they didn't need to, because their cousins, microglia, were already policing the neighborhood.

Microglia, scientists now understood, were functioning like the white blood cells of the brain.

Stevens began looking at microglia more carefully. These tiny cells were breathtaking to watch up close.

Beneath the high-resolution microscope, individual microglia resembled elegant tree branches with many slender limbs. Their branches swirled around and around through the brain, exploring and searching for the slightest sign of distress. As they passed by neurons, microglia extended and retracted their tiny armlike protrusions, tapping on each neuron as if to inquire, *Are we good here? All okay? Or not okay?*—as a doctor might palpate a patient's abdomen, or check reflexes by tapping on knees and elbows.

And microglia were so quick about it. "I'd never seen another cell move so purposefully," Beth recalls. "Not only do they make up ten percent of our brain cells, they are constantly surveying every area of the brain as you and I are speaking," she explains. "If readers find themselves intrigued by what they are reading right this second, microglia are moving around even more! Their day job is surveying our brain. Imagine these little cells everywhere in your brain, just checking things out. How is that neuron doing? How is that synapse doing? How's that circuit? Oh shoot, something is going on there—let's head over and see what's happening!"

Stevens was mesmerized by these tiny dancers. "No other cell in your brain can move around, detect minute changes and respond—and that fact alone was fascinating to me. And yet it turns out microglia were *built* for that."

---

whereas other glia have entirely different origins and are actually part of the nervous system. Indeed, says McCarthy, "Microglia are not really glia. They are not nerve cells, they are immune cells. They are a third entity—part of the immune system."

In Ben Barres's lab, Stevens was already working on a novel project examining how synapses originally come to be pruned and sculpted to form a healthy brain during normal development. Specifically, she and Barres were trying to figure out what role a particular immune molecule known as *complement* might play in pruning away synapses in the developing brain.

At that time, scientists knew that complement plays an incredibly important role in the body. When a cell dies in one of your organs, or there is a pathogen, foreign substance, or microorganism that shouldn't be in the body, complement molecules quickly tag it for removal. And then immune cells—specifically, a type of white blood cell known as *macrophages* (Greek for "big eaters")—see the tag, engulf the cell or pathogen, and cart it away.

In the body, these macrophages also play a big role in inflammation as well as in physical disease, especially autoimmune disease. When activated, they can spurt forth a slew of inflammatory chemicals that cause significant damage. For example, in autoimmune disease, they sometimes mistakenly go too far in their effort to destroy pathogens and begin to do harm to tissue. We see this in diseases like rheumatoid arthritis, in which macrophage immune cells destroy cartilage.

However, complement wasn't thought to play a role in normal brain health and development. The prevailing view of medicine held that the brain was not an immune organ, and therefore there were unlikely to be immune cells akin to macrophages functioning within the brain. So Stevens and Barres weren't sure what could be causing synapses to disappear.

Nevertheless, the two scientists hypothesized that perhaps in some way that they didn't yet understand, complement was playing a role in determining why some synapses in the brain were removed during normal development, while other synapses remained.

As a fetus grows in the womb, the developing brain produces many more synapses than are needed. Extra synapses have to be pruned away to achieve the precise synaptic connectivity that allows for the complex workings of the human mind. During this process of synaptic pruning, some subsets of synapses are eliminated, while others are preserved and even strengthened. As in horticulture, this pruning is a good thing; without it, the brain can't develop properly.

Just imagine a tree that never stops putting out more and more branches, until it's so overwhelmed with growth it can't sustain healthy life. The tree would soon topple over or wither away. The same thing is true of the plethora of synapses that form as a baby is developing in the womb.

Beth Stevens and Ben Barres couldn't help but wonder: What if complement molecules were tagging and sending out "eat me" signals from these excess brain synapses, and these tagged synapses were destroyed — similar to the way in which complement-tagged molecules and tissue were destroyed by macrophages in the body's immune system? What if this was how a healthy, normal brain was successfully modeled during development?

They set out to show — and later proved — that this was indeed the case. When synapses were tagged by complement, these same synapses then sent out an "eat me" signal — and it was these "tagged" synapses that — bing! — vanished from the brain. Think of the way you click and tag emails that you want deleted from your in-box. Your email server's software recognizes those tags, and when you click on the Trash icon, boom, they're gone. That's similar to what Beth Stevens and Ben Barres were seeing happen to brain synapses that were tagged by complement.

Their seminal paper came out in 2007, making a splash in the scientific world. But for Stevens, this discovery answered just one of many questions she had. What were the cause and effect mechanisms at play here? *What was chomping away at tagged synapses, making them disappear?* Was it possible that microglia were involved in this process that so dramatically determined lifelong brain function? Could microglia be the culprit, the macrophage corollary in the brain, responding to "eat me" signals and pruning the brain's circuitry in utero?

And, just as important, Stevens wondered, could this pruning ever go wrong?

Here she began to make one more crucial leap. Beth Stevens wondered: What if this process was not only taking place during development, but was also somehow being mistakenly reactivated later in life, leading to the elimination and destruction of *wanted and necessary* brain circuits and synapses—resulting in diseases years later, even in adulthood? Could it be that these supposedly irritating, do-nothing little microglial cells were actually removing *crucial* synapses in the adult brain? Could an immune cell—one that all of medicine had totally discounted for so long—be deciding what brain circuitry to keep and what should go, thus orchestrating and constantly fine-tuning our brain health?

During the years that Beth Stevens was busy investigating these questions, and on the cusp of discoveries that would cause medical school curricula to be revised, patients, as well as their physicians and psychiatrists, remained completely unaware of these startling scientific advances in our understanding of the human brain.

One of these patients was Katie Harrison, who, in 2008 (the same year that Stevens was wrapping up her postdoc with Ben Barres) was graduating with a Ph.D. in sociology and struggling with devastating psychiatric symptoms that would, in varied measures, derail her life as a young woman, a professional, and a mother.

TWO

"Ten Feet Out of a Forty-Foot Well"

DURING KATIE HARRISON'S LAST YEAR OF GRADUATE SCHOOL, shortly before she received her Ph.D. in sociology, she hit the bottom of the black pit that she had felt herself falling into for much of her young adult life.

Her nerves had become so frayed by the stress of trying to please a difficult Ph.D. adviser that Katie occasionally began to imagine that she was seeing and hearing odd things. One day, when she came home to her small studio apartment after buying groceries, she dropped a bag of baby carrots on the floor. The bag split open and carrots rolled all over the white linoleum. Katie stared down at them, and for a moment she thought she saw orange roaches crawling everywhere.

Around that same time, Katie kept thinking she heard obnoxious music coming from her next-door neighbor's apartment. She went over and knocked on his door and asked, "Hey, can you turn down your music?"

Katie's neighbor told her, "I'm not playing any music!"

The second time Katie knocked on his door he yelled, "Look, stop banging on my door, lady—you're crazy! Get some help!"

Katie, who was first diagnosed with major depressive disorder, or MDD, in college, confided in her therapist. Her therapist explained to

Katie that when depression is not sufficiently treated, patients can experience visual and auditory hallucinations: misfirings of brain circuitry. Katie also went to her psychiatrist, who bumped up the dose of medications that Katie was already on for depression. She also added a mood stabilizer to Katie's pharma-cocktail.

Katie's hallucinations soon abated, but she still felt overwhelmed with anxiety, and she experienced a blurred sense of hopelessness, punctuated by recurring panic events.

It was 2008, and she was thirty-four years old.

Ten years later, Katie Harrison and I are sitting in a café in Arlington, Virginia. We've been shown to a table toward the back of the restaurant, but as we are seated, I notice that Katie's face is white and stiff.

"Are you okay?" I ask her.

"I can't stay in here," she says. "I won't last in here five minutes." The sound of patrons talking, coffee cups clinking against saucers, and waiters bustling is too much for her. "My world has gotten very narrow," Katie explains. "Restaurants aren't part of it."

We walk back out and sit at a quiet table on the restaurant's front patio, away from all the din and bustle. It's a fairly breezy day, but she tells me she'd rather feel chilled than panicked.

Katie begins to tell me a bit about her current life. She's now a single mother with a young daughter and son and she's still struggling with her mood and anxiety. She tells me about a recent morning when she suddenly found herself immobilized by a sense of inescapable dread.

As Katie recalls that day, the dappled light from the trees overhead plays across her face and moppish dishwater-blond curls. Her eyes are dull. It's as if—where small glints of light should shine from her pupils—someone has filled them with brown murky water.

When she woke up on that recent morning, it was raining hard, Katie says. She'd been wide awake all night, worrying about one thing after another until daylight crept through the blinds, promising some modicum of relief. But when she heard the rain, she felt ill; her stomach churned, thinking of all the things that might go wrong in the course of getting her son and daughter off to school.

What if the roads are slick? she thought. *What if the car seat isn't fastened right? If I don't hurry up I'll be late! Then what will the teachers think? I haven't even brushed my hair, I'm so tired . . . if I'm this tired I could have an accident . . . it's safer to just stay home. . . .*

This anxiety came over Katie, as it so often did, swiftly, like a physical illness.

It took every ounce of mental energy she had to get her eight-year-old daughter, Mindy, up, feed her instant oatmeal, get her in her raincoat, usher her out the door, and watch her catch the school bus.

Now she *had* to get her son, Andrew, to school, she told herself. It was Andrew's "Reading Day," a special day when the principal would sit down and read one-on-one with Andrew. He loved Reading Day. Katie could drive him, couldn't she?

Pernicious anxiety was hardly new for Katie. It was often the possibility of loss coupled with the haphazard cruelty of nature that frightened her most. As a teenager, she'd panic when her parents came home late, terrified they'd died. One day in high school health class, when a nurse described how to suck venom out of a snakebite, Katie had fainted out of her chair. At doctor's appointments, blood draws made her vomit or faint. Katie's parents were told their daughter had a "blood injury phobia." By the time she'd landed at an Ivy League college, Katie had developed claustrophobia and took stairs instead of elevators, telling classmates she did it "for the exercise."

Now, as a single, divorced mother, determined to be her best self for her two children, she was trying a new medication cocktail. Katie has also tried cognitive behavioral therapy; talk therapy; EMDR—eye movement desensitization and reprocessing therapy—in which a therapist has her client shift her gaze back and forth quickly while thinking about painful past experiences, in order to dissipate the stress reactions associated with those memories; and somatic experiencing therapy, in which patients learn to focus on the sensations they perceive in their body, tune in to them, and safely observe what feelings arise. She's tried biofeedback and acupuncture, to try to help relax the tight, painful muscles in her shoulders and neck. "My psychiatrist even tested me for neurotransmitter and vitamin deficiencies, and added a half dozen supplements to my regimen," Katie says, in a flat, almost monotone voice. She also sees an integrative physician regularly and was recently diagnosed with Hashimoto's, an autoimmune disease in which the body's immune system mistakenly attacks the thyroid. Katie is trying her best to "self-care." She jogs slowly first thing every morning to get herself "moving," eats a clean paleo diet, and uses meditation tapes to help her sit, breathe, and tame her mind.

And yet, despite all of her efforts, on a morning like the recent one she's describing, at the age of forty-five, Katie felt, she says, that "I had no more control over my anxiety than an epileptic has over their seizures.

"I just stood there, staring out at the downpour through the window in my front door." The raindrops seemed to give the whole world beyond her own threshold an unreal, atomized, impressionistic sheen. "It felt overwhelming to go out that door and into that world," Katie recalls. "I didn't know how I'd drive Andrew, walk him into the preschool classroom, and pick him up afterwards." Still, she tried to press on. She knelt to help Andrew put on his clothes. Her small son stood in the circle of his mother's arms, intermittently glancing at her face as if to reassure himself that his mom was okay. Katie cajoled him out of his footy pajamas and into his trousers.

Then she noticed that the rain was coming down even harder. Dread strained every tired cell of her body and brain. She felt wired, yet overwhelmed by a flattening sense of inertia. That old familiar prickling spread through her like dye spreading through water. It felt as if she'd taken a bad drug. Her fingers began trembling.

"Mommy isn't feeling good," she told Andrew, as she slumped down beside him, her son's shirt still in her hands. "We're going to stay home and have a quiet day today, okay?" She told Andrew he could play on Mommy's iPad for the morning—a special treat—and, for good measure, switched on the TV to PBS.

She didn't even have the wherewithal to call the school to offer an excuse for Andrew's absence without breaking into incoherent, tearful explanations. What reason could she give, anyhow, she remembers thinking: *Hi, I'm anxious and I suffer from depression and I'm on medication and in therapy but today I'm not feeling well enough to drive?*

"I felt so pathetic and ashamed," she says, looking back.

Andrew played on the iPad next to his mom as Katie wept quietly in bed, caught in an undertow of grief for a life inexplicably lost.

She thought back to all the other times she'd let her kids down. A few months earlier, she'd been standing in the pickup line to get Mindy from elementary school, waiting for her third-grade class to get out, when she'd been overwhelmed with self-consciousness. "The other moms were all scheduling playdates," she recalls. It hit her, with the sudden force of a riptide, that she was "the only mom who didn't know a soul." Her life, she says, "had become so limited; all my energy went into just surviving, managing my depression, gutting through my days."

That afternoon, her sense of unease quickly escalated into an all-too-familiar anxiety attack. At moments like those, Katie says, "It's like staring at the sun too long. It's hard to see or think. My mind feels stunned.

"I was dimly aware that a panic attack was coming on, so I walked into an empty classroom and tried to breathe," she recalls.

Another mom, noting Katie's obvious distress, followed her to check on her. "By then I was hyperventilating so badly that the mom ran out of the room, calling for the school nurse. She thought I was having a heart attack."

In her bedroom, on the morning that Katie found herself too unwell to drive Andrew to kindergarten, she slept and rested, then had a neighbor pick her daughter up after school. The next day, Katie felt a little better. Braver. She drove Andrew to preschool. As she walked him into his classroom both teachers came over to greet mother and son, with worried looks on their faces.

"We were so sorry Andrew couldn't be here yesterday for his special day!" the teacher said, hovering beside them. "Is everything okay?"

"What a shame that he had to miss Reading Day!" the teacher's assistant chimed in. "Why couldn't he join us?"

Katie struggled to offer a viable reason. She felt Andrew's warm hand tighten in hers as they stood in the center of the preschool classroom. All around her, the walls were decorated with children's primary-colored handprints and finger paintings: bright smiley faces, shining suns, happy stick figure families holding hands. Her heart began hammering in her veins. "I couldn't get here yesterday!" she finally blurted out.

For a moment, the room fell utterly quiet.

"Both teachers looked at me like I was the world's worst mother," Katie tells me. So did several other moms who overheard the conversation as they were kissing their children goodbye.

Katie kissed Andrew's forehead and fled.

As she drove home through the northern Virginia suburbs, a thousand thoughts tumbled through her mind, Katie tells me. She felt so much shame. But she also felt angry. "How can you describe for people who have never experienced it what it feels like to live inside a panic attack that never really ends?" she asks. I can feel the effort and determination it takes for Katie to choose her words, her sentences, then stop, pause, and give an effortful smile now and then.

She knew, she says, "If I'd said I couldn't call because I'd broken my wrist, they would have asked me, 'How can we help out?' But if I'd been honest and told the truth about my depression and anxiety, the teachers and other moms would have decided I was crazy."

And, Katie confesses, "I made the decision a long time ago that I'd rather have people judge me as 'not being a very good mom' than see me as mentally unfit. I'm more comfortable with that because I *know* in my heart that I'm a good mom—most days I am, I devote all the energy I have to my kids' happiness—but at this point I don't really know. What if I *am* crazy, and other people find out?"

Katie's Ph.D. in sociology and her later degree in social work—she has worked as a counselor at a community health center—have given her an insider's perspective on how patients with intractable mental health struggles are often viewed. Katie knows she would be seen as one of those patients who have had what the mental health field often refers to as "a poor outcome."

Not that she hasn't gotten better. She has improved over the past decade. A little. The hundreds of hours of talk therapy and myriad antidepressants and supplements she's been prescribed have helped her. There have been good days, good months. At times, she's even felt well enough to volunteer at school. "On those good days, I just try to drink it all in, sitting there, working on an art project with my son or daughter, molding clay into dinosaurs, listening to the sound of the kids' laughter."

Katie feels, deep inside, that she is capable of living *that* life. "But I'm just not sure what else to try," she says, with a self-deprecating laugh.

Her decade of hard effort at healing has, she admits, "allowed me to climb ten feet out of a forty-foot well—but that doesn't mean I'm out of the well. And that makes me wonder, what is going on in *my* brain that, despite exhausting every medical and mind-body avenue I can, I cannot climb out of this abyss?" There are tears in her eyes. She wipes at them almost angrily before they roll down her cheeks.

I am reminded of words Vincent van Gogh wrote to describe his depression to his brother: "One feels as if one were lying bound hand and foot at the bottom of a deep, dark well, utterly helpless."

A Nameless Dread

It has always been a conundrum for those suffering from mental health disorders to explain the depth of their pain to those who've never been afflicted. And the difficulty that the medical community has had in understanding mental suffering from a biological perspective has done little to ease that burden.

The word "melancholia" first appeared in English as early as the year 1303. In medieval times, melancholy was said to be caused by an excess of black bile, a bodily humor thought to be secreted by the body's organs.

In the nineteenth century, Victorian physicians began to refer to anxiety and depression as "neurasthenia" in men or "hysteria" in women. Neurasthenia was marked by physical aches and pains, as well as nonhysterical anxieties. Hysteria was marked by bizarre, dramatic, unsettling and theatrical behavior. The best remedy for either affliction* was thought to be a rest cure, during which a patient avoided any physical or mental exercise.

By the late 1800s, melancholy was often referred to as "brain fever"—a period of time when a patient simply couldn't function due to nervousness, or in the aftermath of a traumatic event.†

Then, about a hundred years ago, a Swiss-born psychiatrist at Johns Hopkins came up with the term "depression." The word stuck—though, as William Styron writes in his 1990 memoir, *Darkness Visible*, "depression" does seem "a true wimp of a word for such a major illness . . . leaving little trace of its intrinsic malevolence and preventing, by its very insipidity, a general awareness of the horrible intensity of the disease when out of control."

Fast-forward to the time Katie was in grad school, and anxiety and depression were now decidedly seen as neurochemical—serotonin and dopamine—disorders. SSRIs like Prozac, Paxil, Zoloft, and Lexapro hogged the spotlight in psychiatry. Treatment revolved around finding the right pharmaceuticals for any given patient's chemically deficient brain, and the brain became a laboratory for a psychiatrist's well-intended

---

* The term "hysteria" derived from the Greek word for the uterus: *hysterika*. Physicians in ancient Greece, including Hippocrates, believed that when women exhibited too much emotion or excitement or had female complaints or maladies, it was because they suffered from a "wandering uterus." Their uterus was thought to be literally roaming around in their torso, causing havoc and symptoms wherever it went. If the uterus traveled upward, a woman felt a sluggish torpor (what we might describe as depression). If it moved downward, a woman suffered from a loss of "speech and sensibility" (what we might describe as a panic attack). Oh—and the cure for a "wandering uterus" included prescribing a woman to have more sex with her husband.

† Throughout Sir Arthur Conan Doyle's Sherlock Holmes series, he describes patients suffering from what we would now consider anxiety or depression as having "brain fever." At the time, Conan Doyle had no idea how uncannily prescient this term was.

experiments with what drug cocktail might help that patient best. If necessary, meds were added to help counter antidepressants' and mood stabilizers' side effects: fatigue, weight gain, sluggishness, brain fog.

In 2013, Thomas Insel, then director of the National Institute of Mental Health (NIMH), announced a paradigm shift and policy change in how clinicians should approach and treat anxiety, depression, and mood disorders. Given advances in neuroscience, he posited, it was clear that mental health disorders were biological disorders stemming from changes to brain circuitry and neural structure. According to NIMH's new policy, figuring out how to treat the circuitry differences underlying mental disorders should be clinicians' primary treatment target. Research was clearly showing that in many different disorders of the brain, certain brain circuits weren't functioning and connecting the way they should. Some synapses were offline, and others were terribly overactive: over-functioning.*

Forward-thinking as this change was at the time, it, alas, didn't help patients a great deal, since no one understood why, in diseases like depression, anxiety, obsessive-compulsive disorder, and bipolar disorder, brain circuitry had become so dramatically altered in the first place.

The Isolated Patient

I ask Katie what support she, as a single mom, has from her extended family. "Are they there for you?" She smiles but the corners of her mouth remain turned down. "My family doesn't view having a mental health problem as being totally legitimate," she explains. "If you talk about feeling depressed or anxious, you're being dramatic."

On the other hand, Katie's family often swap stories about their physical ailments, symptoms, and latest medical treatments. Everyone has

---

* Insel went so far as to say that NIMH would move away from the psychiatric categories delineated in the *Diagnostic and Statistical Manual*—and psychiatrists should too—because, when looking at brain circuitry, practitioners found that many of these disorders appeared to be more similar than previously imagined, despite the very different labels psychiatry may have given them.

something. Katie's mom, Genna, who is in her sixties, battles two auto-immune disorders—connective tissue disease, in which the immune system attacks and inflames the connective tissue between joints, and psoriasis, a skin disorder in which the immune system attacks the skin, causing painful, itchy lesions. And her mother's younger brother, Paul—in his late fifties—suffers from crossover type 1 / type 2 diabetes.

Katie's grandmother Alice—her mom's mom, who recently passed away—faced autoimmune and brain-based issues too: Crohn's disease, an autoimmune disease in which the immune system attacks the lining of the intestines, OCD, and Alzheimer's, which Alice was diagnosed with when she was in her sixties.

At family gatherings, Katie tells me, her relatives freely swap medical war stories about their efforts to manage their physical autoimmune conditions—what doctors they're seeing, what medications are working for them. "In my family, physical illness is legitimate; it's something that happens *to* you, and as a medical patient you deserve everyone's empathy," Katie explains. Mental illness, on the other hand, "that's viewed more as a personal weakness, it's about *your* inability to function well. If you can't function well mentally, that's your responsibility." And, she says, "Since I am the one who has the worst brain-related disorder in my family, I am the biggest failure of all."

Much as they might not like to discuss them, however, Katie is not the only one in her family to suffer from anxiety and other mental health concerns.

"When I was younger, my mom also suffered from depression," Katie says. "She's been on and off antidepressants, as I have. When I was a kid my dad would get so upset with her because she wouldn't brush her hair, or take a shower, or brush my hair, and he would say there was something wrong with her. We never talked about it except to ask each other, 'How's Mom doing today?'"

Katie's uncle Paul has obsessive-compulsive disorder (OCD), and her cousin Carly, who Katie says "is like a sister to me" and is six years younger, has been diagnosed with ADD and generalized anxiety disorder

(GAD). She doesn't experience Katie's panic attacks or mind-numbing depression, but she feels anxious about almost everything.

Still, despite the plethora of mental health concerns in Katie's family, it is hard for them to empathize with how challenging life is for her. "If feelings of anxiety were helium, my mom and my cousin would have higher voices, but I would just float away," she says. "Their experience is different, so their ability to understand is limited."

Katie shares a memory of being with her family a few years ago. "We were sitting around the lunch table at my parents' house in Virginia. My uncle Paul, who is really overweight, was telling us about his recent diabetes diagnosis, and the foods he could and couldn't eat. Then he asked if we wanted to see the device he'd been given for testing his blood sugar. He pulled it out of his pocket and started to jab the retractable needle at his finger."

Katie told everyone that she didn't want to see the blood—they were aware she had fainted at the sight of needles all her life—but her parents told her to just stop it. "Oh, come on! No more of this silly stuff!" her mom said, as her uncle demonstrated how he pricked his finger.

"So here I am taking on the shame and blame for my anxiety, while my uncle is proudly showing off his device for managing what is in some ways a preventable disease," Katie says. Her laugh is sudden and loud and uncomfortable. "And yet there was no sense that my uncle should feel any degree of personal responsibility for his illness, even though he'd lived on high-carb and sugary foods and been overweight and never exercised for forty years before he became ill.

"But we don't openly address any of our family's struggles with depression or anxiety. Or with ADD, or OCD or Alzheimer's," she goes on. "I think that's why my cousin Carly hides what she goes through, she never even discusses it.

"Maybe I'm not being entirely fair," she adds, more softly. "I am close to my mom, she would do anything for me. And my parents have helped me—babysitting, paying for treatments, grocery shopping. But they don't seem to grasp how hard life is for me each day. That leads them to expect me to be able to do more for myself than I can do, which makes me feel bad about myself, like I'm somehow failing."

Katie says that they also pitched in to help her grandmother "as best we were able to, especially toward the end of her life. We are a close-knit family in that way." But when it comes to her own mental health symptoms, "I have to turn down the dial of self-disclosure. It can be very isolating."

I feel a lump in my throat—grief for Katie and patients like her. I have heard so many similar stories of mental and physical suffering from so many readers, as they swim against a seemingly impossible tide in their efforts to achieve a healthy, joyful life. It has been said that we, as a society, have decided whose suffering matters and whose doesn't. And that artificial determination begets more suffering. It pains me that for so long the medical and psychiatric communities have had so few answers to offer.

I reach out across the table and lightly cup Katie's hand. I tell her that her family is not that different from many other families in the way they treat physical and mental health disorders differently. And part of the reason for this, I believe, is because her family, like most families, sees their struggles with physical disease as completely unrelated to the brain-based and mental health disorders that also afflict them.

But that is, the new science on microglia tells us, entirely wrong.

THREE

Friendly Fire in the Brain

ETH STEVENS WAS THIRTY-SEVEN YEARS OLD AND WRAPPING UP
her postdoc at Stanford in 2008 when she and Rob packed up their
one-year-old daughter and all of their belongings again and moved back
east in order for her to take up her current faculty position at Boston
Children's Hospital and Harvard Medical School. Rob, meanwhile, had
been offered a job in communications at Boston Children's; they felt
lucky to land positions at the same institution.

On the one hand, it felt like heading home. On the other hand, while
only forty minutes from her childhood backyard in Brockton, Boston was
another world, and heading up her own Harvard lab was a dream that
young Beth could not have conceived of.

The first time Beth arrived at her lab at the newly created Center for
Life Sciences Building at Boston Children's Hospital, she says, "The
space was huge, brand-new, and completely empty. The people I'd hired
hadn't arrived yet. I just stood there alone, totally amazed that this was
ours." Still, Beth adds, "I had to figure out what was I going to *do* with all
this." Already an obsessive observer of microglial cells, she knew one
thing for certain: "I wanted to find out exactly what microglia were up to
in the brain, in a way that would help us better understand human suffering."

Beth had hired just one postdoc, Dori Schafer, whom she'd wooed away, she says, "from the big dogs at Harvard. I was lucky to have her." Beth had also hired one of her grad students and a lab tech. "It was just the four of us and that was it."

When Beth started her lab she didn't know for sure if microglia were eating brain synapses. But she had good reason to believe they were: She knew that microglia were not just passively hanging around waiting for neurons to die and then carting them off like trash disposal workers. Microglia continuously surveyed the brain for even slight disruptions in normal brain activity.

Stevens and others had already observed that when microglia sensed even the smallest damage or change to a neuron, they took swift and immediate offensive action. They headed, spider-like, in that neuron's direction, then drew in their many treelike limbs and quickly morphed into small, amoeba-like blobs. Soon after, those same synapses disappeared. Poof. Gone.

Was it possible that these seemingly insignificant cells were actually blasting apart and engulfing the synapses in question—and then eating them up?

"We could see that microglia were so dynamic, their protruding arms were actively touching synapses, checking them, rushing to the site of injuries," Beth says. "But we didn't know if microglia *directly* interacted with developing synapses or caused them to vanish. No one had ever asked that question."

Just a year earlier, Beth had shown, with Ben Barres at Stanford, that when synapses in the brain are tagged by the immune molecule called complement, those synapses disappear. Were microglial cells able to recognize synapses that complement had tagged—and did they then remove those same synapses?

If Beth's theory was right, this would mean that, in early development, microglia were responsible for sculpting the human brain. But Beth was still fixated on the same question she'd first posed while at Stanford. What if something else was going on too? "What if, later, in the teen years, or even in adulthood, the same developmental pruning gets mistakenly turned back on again? Only now it's a bad thing?"

If microglia were mistakenly eating away synapses that *shouldn't* be pruned, this would be similar to the way in which certain types of white blood cells—which serve as our immune system's first line of defense—behave in physical health disorders. Remember, in the body, when the immune system detects an outside threat—an infection, environmental chemical, virus, foreign pathogen, physical trauma, or chronic emotional upheaval, which floods the body with stress neurochemicals—white blood cells morph and change into blob-like macrophages, which set out to find and eliminate any foreign invaders. But sometimes when the immune system goes into overdrive, it doesn't know when to *stop* causing inflammation or destroying cells—and it can do a lot of collateral damage, as in conditions like Katie's thyroid disease, Katie's mom's connective tissue disease and psoriasis, and Katie's uncle Paul's crossover type 1 / type 2 diabetes. This is also the case in diseases like lupus, scleroderma, multiple sclerosis—or the disease I've twice battled, Guillain-Barré syndrome.

It's a case of friendly fire: the immune system running amok.

And that's what intrigued Beth Stevens most. If microglia, which researchers now understood were akin to the macrophages of the brain, were altering brain circuitry, perhaps, like white blood cells, microglia didn't always get it right. What if, instead of just pruning away damaged or old neurons, microglia were sometimes mistakenly engulfing and destroying healthy brain synapses too?

"Diseases like schizophrenia and Alzheimer's and autism are completely different in respect to their timing, their genetic predispositions, the parts of the brain they impact," Beth explains. "But could it be that there was a common disease pathway and maybe that common denominator was that microglia were causing synapse loss?"

Increasingly, disorders of the brain, from depression to learning disorders, were being understood as circuitry disorders; certain synaptic connections in the brain were not firing and wiring together as they should.

"What if these little overlooked glial cells we were setting out to study were at the center of it all?" says Beth. Her hands spread wide with excitement as she explains. "You can imagine how if you have too much pruning or too little pruning, things go awry! You could have too many synapses, or not enough synapse connectivity. And you can imagine, given how our brain works, if that connectivity is even slightly off, that could potentially underlie a range of neuropsychiatric and neurodevelopmental disorders as well as cognitive impairment."

Could it be that sometimes microglia were the brain's sculptors and protectors, the angels of the brain, and sometimes they were the brain's untimely assassins?

Was it possible that scientists had, for so many decades, missed something so huge?

No one knew.

If Beth could prove this grand, complex hypothesis, it would completely change our understanding of brain health, from womb to grave.

But in order to prove her hypothesis, she first had to get a better look at these little cells.

Back when she'd done her postdoc in Ben Barres's lab, the Stanford team had used a model of the brain's visual system—the optic nerve and retina—to show that synapses tagged by complement disappeared. Such synapse loss in the retina was known to lead to diseases such as macular degeneration, glaucoma, and blindness. So Beth stuck with using visual systems, in animal models, in order to get a better look at what microglia were up to.

She hypothesized that if, indeed, microglia were engulfing and destroying synapses, then you would see chewed-up synapses *inside* the microglia. "That was the big question," Beth explains. "Could we find pieces of synapses *inside* the microglia? Dori, my postdoc, figured out a really clever way to allow us to see if that was indeed the case."

Dori Schafer, Ph.D., now an assistant professor of neurobiology at the University of Massachusetts Medical School and Brudnik Neuropsychiatric Research Institute, explains their experiment—one that is now universally considered a landmark study in the field of neuroscience.

In order to get a closer look at how microglia were interacting with synapses, Dori injected dye into the eyes of mice.* This dye then traced down from the neurons in the eye nerves and deep into the brain. This, says Dori, made the brain's synapses "glow bright fluorescent red." The microglia, on the other hand, were stained fluorescent green (which seems to be microglia's signature color in experiments). "And this allowed us to clearly see both the synapses and the microglia, since each glowed a distinct fluorescent color."

This process—simply finding a way to accurately see the interplay of synapses and microglia inside the brain—took nearly a year.

"One weekend I was in the lab by myself, imaging and reimaging microglia and synapses," Dori recalls. "I'd looked through the microscope a million times. Then, suddenly, I saw these red structures—the synapses—glowing like fluorescent red lit-up little dots, and these red dots were *inside the belly of the green microglia.*" Dori was stunned. "I kept thinking, *We were right! Microglia* are *eating synapses!* I was staring at proof positive with my own eyes," she says.

Dori didn't immediately tell Beth what she'd found—not yet. "I wanted to be sure. So I repeated the experiment a few more times that weekend. And there it was, every time: The synapses were inside the microglia. Microglia were chomping them to pieces."

Beth recalls Dori running into her office that Monday morning with photo images in her hand. "They're in there!" Dori told Beth. "The synapses are really inside the microglia! We can see it!"

"It was such a high-five moment," Beth recalls. "Microglia *were* like tiny little Pac-Men in the brain—and brain synapses were in the bellies of the Pac-Men! We were on to something really wonderful, really novel. This was deeply important in terms of looking ahead to microglia's role in disease."

* Here, as in all stories in this book that involve animal studies, let's say a silent "We're sorry" and "Thank you" to the mice; without such lab research, we would know very little about the human brain, or how to help people who are suffering, since it would be highly unethical to open the brains of living human beings to observe what's going on in real time.

· · ·

Beth and Dori rolled up their sleeves. "We were both really fired up—but it was also really hard and stressful," Beth recalls. She was pregnant with her daughter Zoe and had a toddler at home. "Dori and I knew we had to really crank. We had so much adrenaline, but we also wanted to be very rigorous and take our time doing our control experiments and data analysis. We wanted to do it right *and* publish it before anyone else."

It was a tall order.

The following weeks and months were reminiscent of Beth's early days as a scientist, back when she'd started out in Doug Fields's lab. Many nights she'd look up at the clock and it would be close to midnight. "It just wouldn't make sense to go home," she says. But one thing had changed compared to her early days as a lab tech. Beth no longer had to sleep on a pile of clothes under the lab conference table. Laughing, she says, "Dori gave me a blow-up mattress as a present. I put it under my desk and if it was really late, I'd just crash there."

Beth's husband, Rob, she says, "helped make everything possible. He knew the importance of what we were doing. He helped Riley understand that if I wasn't there to put her to bed some nights it was because I was trying to do work that we believed would help a lot of people."

In 2011, Beth and Dori submitted a paper outlining their findings for publication and review. Beth had recently given birth to her second daughter, Zoe, and now had two young children at home. And Dori had recently gotten married.

It was a watershed year. In 2012, their seminal paper was published in the journal *Neuron*. It was the first scientific study offering evidence that complement was sending microglia "eat me" signals, and that microglia were pruning and eating away developing synapses. Microglia, they showed irrefutably, held the power to engulf and remodel *healthy* synapses.

Not surprisingly, the scientific world exploded with the news. Their study was later cited as the journal's most influential paper that year.

Meanwhile, researchers at the European Molecular Biology Laboratory in Italy showed that microglia could be especially overactive in the brain's hippocampus—an area crucial to mood and memory. By engulfing and removing healthy synapses in the hippocampus, microglia were causing a loss of brain matter in a part of the brain that was known to be deeply affected in depression, anxiety disorders, autism, obsessive-compulsive disorder, and Alzheimer's disease. In these diseases, the brain's hippocampus had been shown—through PET scan imagery—to be visibly atrophied.

These revelations solved a decades-old mystery. In many different neuropsychiatric and neurodegenerative diseases of the brain, healthy synapses disappeared; neurons died off in droves. But no one had ever understood why.

Suddenly it all made complete sense.

Even though microglia were trying to protect the brain and make it healthy and strong, just like white blood cells in the body, when microglia detected that something was off—an overinflux of stress hormones, an infiltrating virus, chemical, allergen, or pathogen on the scene—they all too often went too far, removing every synapse hanging out in the neighborhood.

And that understanding—that this little boring cell that had been hiding in plain sight was an immensely active, and sometimes hyperactive, immune cell—changed everything.

In 2015, Beth Stevens was awarded a MacArthur "genius grant" for her discovery of the role of microglial cells in synaptic pruning during development and disease.

One Cell with Many Faces

Thus far we've been focusing mostly on microglia's dark side.

But these tiny little cells have a bright side too. When the brain is in a state of homeostasis—in other words, when microglia aren't being triggered to misbehave—they get active in a completely different, positive way. In a healthy brain, microglia secrete nutrients to stimulate new, healthy neurons to grow and create brand-new synapses, wherever they might be needed, to help mend the brain. They even release neuroprotective factors involved in repairing sick neurons.

In fact, microglia can directly help neurons to reach out and form new neuronal projections—a little like growing a new appendage—and these new growths can then clamp on to other neurons, thus increasing and strengthening brain connectivity.

Microglia, along with other types of glial cells, also foster the growth of myelin, which insulates fibers in the brain, helping to speed up synaptic connections. And one of the most active areas where microglia do such repair work is in the brain's hippocampus.

"Microglia have so many good roles if they are balanced just right," Beth underscores. "When they are in a state of homeostasis, they release signals that spew out all these different proteins and good chemicals that are protective. They actually try to stop the process of synapse loss."

But the minute that microglia perceive a change, or somewhere that something tiny goes wrong, or a big thing goes wrong, they can stop emitting good, protective neurochemicals and spit out neuroinflammatory chemicals that harm the brain. And that can cause another kind of damage, in addition to synapse loss: runaway inflammation. "If something shifts, microglia can get pushed into a pro-inflammatory state and begin releasing a lot of cytokines, making them one of the biggest producers of inflammatory chemicals in the brain," Beth explains.

For instance, if there is a traumatic brain injury, "microglia go nuts," she explains. "They can start cranking out inflammatory signals that may initially help protect the brain but can become dysfunctional and start to produce signals that activate other glial cells, like astrocytes, to release toxic factors and harm neurons."

At the University of Maryland School of Medicine, Margaret McCarthy, Ph.D., a professor of neuroscience, focuses on microglia in brains much younger than those studied by Beth Stevens. McCarthy has found that microglia can be programmed by early experiences—exposure to hormones, infection, or inflammation—in ways that change how well microglia respond to later events in life, including future trauma, stressors, or infections.

Scientists now hypothesize that once microglia are triggered in the brain, this may cause changes to the genes that oversee how microglia will behave over the long term, reprogramming them to be set on super-vigilant high alert—and making them even more likely to be hair-triggered to overprune synapses or misbehave in the future.

Set on overdrive, microglia don't stop pruning away brain connections when a threat has passed. They can keep on spewing out inflammatory chemicals and taking down synapses *even when the stressor or pathogen is no longer there.* Neuroinflammation becomes a self-sustaining, run-away process. And this runaway process can translate into changes in the brain, even years after the original inflammatory process began. Some-thing that affected the behavior of microglia in the brain early in life may appear as an anxiety or behavioral disorder, or depression or schizophre-nia, in the teenage years, or as Alzheimer's late in adulthood.

Beth Stevens hypothesizes that it could be that initially it's just a small environmental hit that changes the brain—an infection, environmental toxin, trauma, physical or emotional abuse, chronic mental stress. And the brain is coping. "But then comes along another environmental hit—and together it creates a perfect storm, and suddenly we end up with something really bad."

In psychiatry, physicians still don't know why mental health disorders become so intractable in one person versus another. Runaway microglial inflammation, triggered by environmental exposures, life experiences, and genetics—a combination that is unique to each individual—might offer us a clue to the answer. When microglia go on a full-throttle attack, they take out crucial synapses in the brain that we need to process thoughts, manage complex emotions, and make good decisions. We may feel it keenly. Important parts of the brain that should be talking to each other can't communicate well. Synapses misfire. It's hard to make sense of the world around us. Perhaps when something seemingly small happens we overreact. We feel despair. We can't concentrate. We act out. We may feel elated one moment and devastated the next. Perhaps we can't remember things. Or we feel anxious all the time. Or some combination of the above—it's a little different for everyone. And so we give it a hundred different names. Learning disabilities, OCD, ADHD, anxiety, depression, bipolar disorder, postconcussion syndrome, you name it.

But what if we can look at the world in a slightly different way? And, instead of asking "Why am I feeling this way?" or "Why can't I get a grip on myself?" or "Why am I forgetting things?" we start to ask, "Why are microglia taking out synapses and making me feel this way, and what can I do differently so that this process stops happening?"

Beth Stevens is on a mission to help figure out what makes microglia start eating synapses and spew out inflammatory chemicals in the first place. If she can ferret out the biological cascade that turns good microglia into bad microglia, she can also figure out how to get them to shift back again.

But first, for the sake of scientific precision, Stevens wanted to test the hypothesis that microglia were causing synapse loss in a range of different specific brain-related diseases. It was one thing to show, in general terms, that microglia were capable of eating synapses. But it was another thing to prove that microglia were responsible for the changes in the brain that led to specific diseases such as Alzheimer's, schizophrenia, and autism.

In 2016—with generous foundation funding—Stevens and her team collaborated with Ben Barres's lab at Stanford and showed that overhungry microglia *were* contributing to the early loss of synapses in Alzheimer's disease animal models. In Alzheimer's, abnormally high levels of the molecule complement tagged too many brain synapses, setting microglia on destructive mode, leading to the loss of otherwise functional brain circuitry.

And this synapse loss in Alzheimer's, it turned out, was starting at a remarkably early stage in the disease. Stevens and her collaborators showed that microglia were destroying healthy synapses in the brain, including in the hippocampus, long before amyloid plaques formed in the brain and neuroinflammation occurred.

And this revelation—that aberrant synapse loss occurred far in advance of significant Alzheimer's symptoms—led Beth to consider two more possibilities.

First, if synaptic changes were occurring so early in disease, could it be that microglia were sometimes engulfing and destroying synapses that were less active—that is, neural circuits that were demonstrating less than ideal neuronal activity? In other words, were microglia eliminating these less active neural circuits simply because they weren't firing up a lot?

Beth hypothesized that in schizophrenia and other neuropsychiatric diseases, neural pruning might also be happening years before disease symptoms appeared. These synapses might be targeted for a host of reasons: stress chemicals, pathogens, foreign invaders—or a simple lack of activity.

So Beth turned her attention to tackling psychiatric diseases. She wanted to test the concept that microglia were inappropriately pruning synapses in the prefrontal cortex area of the brain during critical stages of development, such as adolescence. This aligned with the work of other researchers, who'd shown that the number of neural connections in the prefrontal cortex appear to be fewer in individuals with schizophrenia than in unaffected people.*

However, there were no good animal models to test this idea.

In 2016, a colleague of Beth's, the geneticist Steven McCarroll, Ph.D., discovered that higher levels of complement were associated with a significantly higher risk of inappropriate pruning, and with schizophrenia. This sparked a collaboration between Stevens, McCarroll, and another colleague, the Harvard immunologist Michael Carroll, Ph.D., to connect the dots between this genetic discovery and the pruning mechanism that Stevens had been investigating in mice.

Were microglia once again the culprits, gobbling up complement-tagged synapses? It seemed very possible that this was the case. "Imagine if we can show that in neuropsychiatric diseases, this synapse loss is oc-

---

* This theory, known as the Pruning Hypothesis, holds that symptoms in schizophrenia and diseases associated with it, such as bipolar disorder, result from too much synaptic pruning at vulnerable windows in brain development, such as adolescence and young adulthood.

curring very early on, and we could *know* when synapse loss first starts?" Beth asks, her voice full of optimism.

We might, she continues, "be able to peer inside the brain of an other-wise healthy-seeming adolescent, see that too many synapses are disappear-ing at the age of ten, or twelve, and help them years before psychiatric symptoms appear," perhaps even preventing the disease from occurring.

Or, she goes on, "Let's take Alzheimer's. Wouldn't *you* want to know if you were losing synapses twenty years before you developed symptoms?"

"Indeed," I say, "I would."

"*I'd* want to know if I was losing synapses so that I could take whatever steps to help keep those synapses from disappearing!" she says, excitedly. "What if we could catch these very early stages of synapse loss—you know, the kind of thing that makes us forget where our keys are, long before we forget who our mother is?"

I think of patients like Katie, who already suffers from several psychi-atric disorders, and whose grandmother Alice faced early Alzheimer's.

In 2018, Beth was named a Howard Hughes Medical Institute Investiga-tor, one of science's most prestigious honors, and awarded $20 million to follow the bread crumbs and further learn how microglia might be con-tributing to a range of diseases that tear lives asunder, and how we might stop that from happening.

Behind Beth's desk sits a shelf, and on that shelf, amid photos, coffee mugs, and awards, there is a set of four handmade beer glasses that her lab team had crafted for her—to better enjoy the "microgliale" beer they brew right here in her lab. Etched onto every glass are these words, in elegant script: *Microgliale: Engulf Responsibly.*

And that is the big question: Can scientists and clinicians find ways to make sure microglia engulf only the synapses and circuitry in the brain that we don't need—so that we can retain all the synapses we do need in order to be healthy and happy throughout the many different seasons that make up a lifetime?

Microglia Everywhere

L ILA SHEN—MY FRIEND WHO ONCE MISTAKENLY LEFT HER TOD-
dler behind in the kindergarten classroom after dropping off her
older son at school one morning—and I are headed out to walk and talk
at a local park. Lila is in her early fifties and has dark hair through which
there runs a single long, elegant streak of silver, jetting down from her
left temple and over her shoulder. After a decade of living with Crohn's
disease, ADD, and obsessive-compulsive disorder, Lila is still struggling
to find a manageable status quo. The rhythm of her life is occasionally
punctuated by sudden hospital stays, and some days she feels so scattered
and forgetful that, she says, "It's getting a little frightening."

"I can get by like this," she tells me, as we pass a group of joggers—
something neither of us can do given the limitations imposed by the
autoimmune conditions we manage. "But what will happen to me if I get
even more forgetful over the next ten years?" This worry crosses her
mind every day—each time she fails to recall something important.

Still, that's not Lila's most immediate concern. What's really worrying
her at this very moment, she says, "is the toll that all of my problems have
taken on my boys." She worries that her chronic health problems have
robbed them of their childhood and eroded their sense of well-being.

Lila's younger son, Jason, is now eleven. From the time he was a toddler, he's watched his mother disappear every so often to spend several nights in the hospital for treatment for her inflammatory bowel disease. She's frequently forgetful about events on the calendar, leading to missed baseball practices and overdue school forms. She gets lost a lot while driving. And, she says, "I'm obsessive about things like repeatedly cleaning the countertops and washing our hands." Lila doesn't eat in restaurants because "a single bout of food poisoning can mean I lose a few weeks of my life. The kids have grown up in a house full of worries."

Jason was much younger than his older brother, Liam, when their mom's health issues first became so pressing, and therefore, Lila thinks, he was more deeply affected by his mom's medical trauma, because he didn't understand why she was sometimes unable to do things with him.

And now Jason, she's noticed, is showing distinct signs of his own anxiety issues. Just a few months earlier, while attending an annual bayside camp where campers swim and sail, the camp counselor, who knows Jason well, took Lila aside and asked her if everything was okay. "The counselor told me that Jason was refusing to go into the water, because he was afraid there might be sea lice, or he might get pinched by a crab." In a way, Lila says, she could understand—the Chesapeake Bay is full of lively critters. But sea lice are extremely rare, and crabs usually scuttle off before you can even get near them. "He was fine at camp every year before now!" Lila says. "He usually loves to play in the waves. He grew up on the shores of the Chesapeake Bay!"

Then, around the time school started, Jason showed other signs of pernicious anxiety. He came home from school one afternoon and said that he wanted to stay home the next day. Lila knew his class was about to give reports on famous writers. "He'd done his on Lois Lowry, who wrote *The Giver*, and it was well thought out," says Lila. "He'd practiced for my husband and me a half dozen times. It was great!"

Lila called Jason's elementary school and asked the teacher if he could help Jason through his anxiety, and she and her husband coached him on deep breathing techniques. Then they sent him to school.

Later, the school nurse called to say that Jason was in her office. He'd started to give his talk, but halfway through he'd stopped because he was so anxious that he was finding it hard to get the words out. Tears had welled up in his eyes. The teacher had taken him out of the classroom to talk with him privately, and Jason had said, "I don't feel good, my stomach hurts, I need to go home!"

Lila took Jason to talk to a therapist, who said he is suffering from anxiety. Lila is duly concerned. "Recognizing my own issues, of course it worries me to see my son facing these issues at such a young age." Lila also realizes, she says, that "anxious parents make for anxious kids. I'd be blind if I didn't realize how witnessing my health issues has influenced how anxious Jason feels in the world."

And that has made her wonder about a larger question, she says. "If a lot of what is going on with Jason is because he is so worried about whether I'm okay, how do I know what is just the psychological effect of that, versus what changes could be happening in his brain?"

She feels in many ways that she is Jason's problem. If his home life were less fraught—meaning, she says, if she weren't struggling with so many health and cognitive issues, and if she were not so anxious all the time—wouldn't he feel less anxious as well? "If I find better answers to treat myself so that I'm not so scattered and forgetful and worried, won't that help Jason feel calmer too?" she asks. "Or has growing up with the uncertainty of my illness changed how his brain works, in some way?"

Lila is both concerned and curious. And she brings up such an important question—one that, it turns out, is on Katie Harrison's mind too. Indeed, a few weeks after my walk with Lila, I find myself having a similar conversation with Katie over Skype. As a single mom, Katie is particularly worried about the effects her depression and panic disorder have had on her kids, Mindy and Andrew.

"I worry that my fear and anxiety have increased their fear and anxiety," Katie says. "Or if they somehow feel that the fact that I'm not well is their fault—that they worry that something they did—being too active, or too noisy—has *made* me sick."

In understanding psychiatric and cognitive disorders, how do we parse out what is situational versus what is biological—for instance, stressful conditions versus overexcited microglia beginning to alter brain synapses in the developing brain? And once we understand how the two—chronic situational stress at home and changes in the brain—are interrelated, how do we utilize that information to better help families like Lila's and Katie's?

Traumatized Microglia

Decades of research tell us that the brain is highly plastic and constantly changing, in an interactive dance with our environment. As Beth Stevens has explained, many different factors in our day-to-day lives can affect the activity of microglia in the brain on a biological level.

One of these environmental factors is chronic stress or emotional trauma. When children face chronic unpredictable stress, and their stress response is routinely set on high alert, the immune system churns out high levels of inflammatory stress chemicals—and this toxic cocktail creates profound changes in the way in which the immune system functions.

When inflammatory stress chemicals routinely flood a child's developing body and brain, this also alters the genes that oversee the stress response. And this in turn can reset the stress response to high—so that, without adequate intervention, the stress response becomes stuck in "fight-or-flight" mode. In fact, Yale researchers recently found that children who'd experienced chronic adversity show changes to the genes that oversee the stress response on all twenty-three chromosomes. This heightened stress response in turn leads to more increased production of high levels of inflammatory chemicals.

And this is the reason why children who grow up facing chronic unpredictable stress are statistically many times more likely to develop a host of physical diseases in adulthood, including autoimmune diseases, heart disease, and cancer. We also know that children who experience a lot of childhood stress are three times more likely to develop depression or other mental health disorders by the time they reach adulthood.

Kids who face chronic stress show significant changes in brain architecture too. When researchers run brain scans on adults who faced multiple types of stress in childhood (often referred to as adverse childhood experiences, or ACEs), which includes the stress of growing up with a parent with chronic medical issues or a mental health disorder, the brain's hippocampus is often smaller, and slightly atrophied, compared to those who didn't experience childhood adversity.

In teens diagnosed with persistent depression, the hippocampus already shows signs of shrinkage as early as adolescence. This means that neurons in the hippocampus are dying off. Our hippocampus is the area of the brain responsible for helping us to respond emotionally, and in appropriate ways, to the world around us, based on our memories, interpretations, and perceptions. It is, in many ways, along with other areas of the brain, where much of our sense of self resides, our understanding of who we are in the world. When the hippocampus's circuitry is over-sculpted, it doesn't just change a person's ability to process memories and emotions, it changes all sorts of behaviors too.

Similarly, in brain scans, children and adolescents who experience chronic stress show fewer neural connections between the brain's hippocampus, the amygdala (an area of the brain that alerts us to danger), the prefrontal cortex (where we make decisions about how to respond appropriately to the world), and the default mode network (an area of the brain that helps interconnect all these other areas and is also correlated with one's sense of self).

This means that kids like Jason may have changes in the synaptic connections in areas of the brain that should tell him whether or not he can feel safe in any given situation, but do not because some circuits are simply not functioning the way they should.

So, although a stressor may be emotional or situational in nature, it can lead both to changes in the architecture of the brain and to a heightened immune response in both body and brain.

We also know that a heightened physical immune response is directly tied to a greater likelihood of developing brain-related disorders. During the same years that Beth Stevens was setting up her Harvard lab, studies in the field of immunopsychiatry were beginning to show a clear link between individuals who had high levels of inflammatory biomarkers in the body and those who had brain-based disorders including depression, learning disabilities, autism, Alzheimer's, obsessive-compulsive disorder, and other mood disorders.

Patients like Katie, who suffer from major depressive disorder, often have much higher levels of the pro-inflammatory cytokines known as interleukin 6 and C-reactive protein—31 percent higher than those without depressive symptoms. And elevated levels of inflammatory chemicals in the body often precede psychiatric symptoms by years. For instance, women with elevated levels of C-reactive protein in 2008 had triple the likelihood of developing depression by 2012, compared with women with lower inflammatory biomarkers. And kids who at the age of ten (just a year younger than Lila's son Jason) had high levels of interleukin 6 and C-reactive protein had a significantly increased chance of developing depression by the age of eighteen.

In 2015, researchers found higher levels of the cytokine known as tumor necrosis factor, or TNF, in the brain's hippocampus in patients suffering from major depressive disorder—and these same patients were more likely to also suffer from chronic pain.

In patients with bipolar disorder, inflammatory biomarkers skyrocket during periods when depressive symptoms worsen and go down significantly when patients go into remission. Neuropsychiatrists have also found relationships between physical inflammatory biomarkers and generalized anxiety disorder (in patients such as Katie's cousin Carly), as well as depression (as in Katie's case). The higher an individual's levels of inflammatory biomarkers, the more prevalent their psychiatric symptoms tend to be. This is also the case in schizophrenia. Moreover, this turns out to be true even when no signs of physical illness or inflammation are detected.

Perhaps most astonishing, in 2017, researchers at Johns Hopkins School of Medicine showed that they could use physical inflammatory markers to predict suicide attempts. Simply activating an individual's inflammatory immune response—as if they're fighting off a viral infection—can trigger feelings of deep despair, and even suicidal thoughts.*

So here's what we know with certainty: When triggers, including

---

* Individuals with autism have elevated inflammatory immune markers, too.

chronic stress, infections, or toxic chemical exposures, routinely spike the body's inflammatory immune response, this can in turn lead to a loss of synaptic connectivity and inflammation in the brain, which can translate into psychiatric, developmental, and cognitive disorders.

Which leads us, of course, to the next volley of questions: What, exactly, is the relationship between elevated inflammatory immune biomarkers, microglial activation, neuroinflammation, and psychiatric disease?

The answers turn out to be stunning.

The Microglial Connection

In 2017, researchers reported that after only five weeks of experiencing chronic unpredictable stress (the kind of stress a child like Jason might feel not knowing from one day to the next if his mother is going to be okay), microglia in the hippocampus in mice begin to show signs of dysfunction. Soon after, significant symptoms of depression appear. This process would take a lot longer in humans, of course, since we have a far longer life span than mice do. However, such evidence suggests that these same changes might occur in the brain over the course of years, or decades, in human beings who face chronic, unpredictable stress.

Indeed, scientists now believe that most molecular pathways to depression are linked through microglia-associated neuroinflammation. And periods of worsened anxiety and depressive symptoms also correlate with heightened microglial dysfunction in the brain. In a recent study in *JAMA Psychiatry*, researchers reported that patients experiencing major depressive episodes have significantly higher levels of activated microglia. And patients with obsessive-compulsive disorder, like Lila, show microglia-activated neuroinflammation with the neurocircuitry associated with OCD.

A similar finding was announced in 2017, when Beth Stevens's collaborator in schizophrenia research Michael Carroll showed that in people with the autoimmune disease lupus, microglia "become reactive and engulf neuronal and synaptic material," causing "microglia-dependent synapse loss" that manifests in the symptoms of psychiatric mood disorders. In lupus, clearly, microglia can become excessively active in their pruning of synapses, triggering psychiatric symptoms.

Suddenly, fifteen years of head-scratching data, which has repeatedly shown that patients with lupus are as much as 75 percent more likely to experience depression, anxiety, cognitive impairment, and in some cases, psychosis, makes absolute scientific sense.

Likewise, patients with MS who report cognitive and memory issues also show a massive infiltration of microglia—which appear to be wreaking destruction—in their gray matter lesions.

Meanwhile, the reason why patients with Crohn's disease, like Lila, have long shown slower response times in cognitive tasks such as solving simple math problems during periods when their disease symptoms flare (as compared to those who don't have Crohn's), is that during flares, Crohn's patients have a much higher level of microglial activation in the brain (as compared to periods when their disease is in remission). Somehow, the body's hyperactive immune system appears to alert microglia, and they in turn become hyperactive—just like their white blood cell neighbors in the body downstairs.

PET scans of the brains of men with autism show a profusion of activated microglia, particularly in the cerebellum—a region associated with processing sensory information, movement, and learning. In individuals with autism, microglia are more perpetually active and numerous and stir up more inflammation.

Microglia also promote disease progression in patients with Parkinson's, and even in West Nile virus, researchers have found that microglia engulf synapses at a dangerous pace, perhaps accounting for the chronic memory problems that more than half of people report experiencing after having had the virus. This helps to explain why, in my own case, after I developed Guillain-Barré syndrome, I experienced such a steep decline in my mood and developed memory glitches.

This can all seem really overwhelming. It's a little frightening to think that our brain is this sensitive to stimuli.

But there is also good—actually great—scientific news pouring in, in equal measure.

A Radically New Way of Looking at Human Well-Being

If anything, the new understanding that microglia can respond to chronic hair-trigger stressors and mount an immune response in the brain helps us make even better sense of how stress and trauma change the brain, thus changing mood and behavior.

Chronic stress is simply one more environmental hit that can knock microglia off kilter, causing them to hypersculpt brain synapses so that essential neural connections are lost. The stress that Lila's and Katie's children might experience from witnessing their moms' chronic symptoms and limitations, combined with whatever familial genetics are at play and any other environmental stressors, may propel microglia to chisel away at synapses in areas of the developing brain that might, over time (and, it should be underscored, without intervention), lead them to perceive the world as a darker and scarier place.

Nevertheless, I reassure both Lila and Katie, the next time I talk with each of them, that this research does not mean that such changes are irreversible. The brain, as we know, is enormously, remarkably, beautifully plastic long into adulthood. And it is especially plastic in childhood.

That means that Lila's and Katie's efforts to turn over every stone they can in their own journeys to find wellness for themselves, and all the steps they take to help Jason and Mindy and Andrew, such as starting talk therapy early on, are crucially important.

"The brain, and microglia, are highly influenced by both the stressors and the positive influences in the environment," I reassure both Lila and Katie. "So all the tools you are implementing to ameliorate stressors help influence both your brain and your children's brains in a good way too."

Simply put, our novel understanding that microglia sculpt the brain, and that we might be able to calm them down or even reboot them so that they stop triggering inflammation and snipping away at synapses and instead protect our gray matter, promises to offer patients like Lila and Katie new tools in their healing toolboxes.

And that's pretty exciting.

Erasing the Line Between "Mental" and "Physical" Suffering

When we look at a family like Katie Harrison's, which faces a range of both physical and mental health disorders, through the lens of microglial immune cells in the brain, we can see that her family's mental and physical health disorders have almost the same origin stories. In some family members, like Katie; her mother, Genna; her uncle Paul; and her grandmother Alice, some combination of inflammatory triggers coupled with genetic predispositions is causing an autoimmune response in the body—announcing itself in Katie's thyroid (Hashimoto's disease), the tissue that surrounds Genna's joints (connective tissue disease), Genna's skin (psoriasis), the beta cells in Paul's pancreas (diabetes), and the lining of their grandmother's intestines (Crohn's disease).

In the case of Katie's major depressive disorder and anxiety, her cousin Carly's anxiety, Genna's depression, Paul's OCD, and Alice's Alzheimer's disease, immune-triggered inflammation has been brewing deep in the brain, as microglia damage synapses, spew out inflammatory chemicals, and destroy neural connections, causing diseases that her family has long found difficult to understand, talk about, quantify, and treat.

And these disorders are also very possibly related to their levels of physical inflammation.

But, I underscore to Katie, that's not always the case. Research into glia also tells us that in some people, like Carly, inflammation and disease might not show up in the body at all—only in the brain. In other words, the same triggers that might cause a single organ or area in the body—such as the pancreas, or the tissue around the joints—to become inflamed could instead cause only the brain to become inflamed.

Except for one major difference.

We know from the work of Beth Stevens that inflammation in the brain doesn't look like inflammation in the body. When the brain becomes inflamed it doesn't become visibly red, hot, painful, and swollen. Instead, microglia promote inflammation by either spitting out inflammatory chemicals that cause damage to important neural structures or engulfing and eliminating synapses so that they disappear altogether.

Just as white blood cells congregate and inflame an area of the body, microglia congregate in an area of the brain—and destroy circuitry. Scientists now refer to this process of excessive glial attacks on neurons and synaptic overpruning as *neuroinflammation* (or, in diseases like Alzheimer's or Parkinson's, as *neurodegeneration*). In disorders like autism, obsessive-compulsive disorder, and mood disorders, neuroscientists call this process *neurodevelopmental changes*.

But whatever term we use to refer to these brain changes, they all mean the same thing: Tiny microglia are engulfing and destroying synapses, and this is the catalyst that sets in motion hundreds of different disorders and diseases that have long remained the black box of psychiatry and neurology.

This means that the long-held line in the sand between mental and physical health simply does not exist. When an individual's immune system is overtaxed, for some, disease may show up in the brain, while for others it may show up in the body. It could inflame your joints, or your mind—or both.

All the big, bad, and difficult-to-treat brain-related diseases of our century, all of which are on the rise, share one common denominator: Immune-triggered microglia are wreaking havoc with the brain—very often in response to the same things that spark inflammation in the body.

Researchers call this new field of inquiry—studying the interrelationship between disorders of our immune system and disorders of the mind—*neuroimmunology*. And science in the neuroimmune field is changing the way we look at the human body, and human suffering, in its entirety.

The field of neuroimmunology helps us to better grasp why a family like Katie's is more likely to face both mental health and autoimmune challenges, and, statistically, why autoimmune diseases appear to be far more common among individuals with mood and cognitive disorders and their immediate families.

Researchers have long assumed there must be a genetic link: Individuals who are more likely to have severe infections or autoimmune disease must also be more likely to have a genetic predisposition for depression, bipolar disorder, and Alzheimer's. And vice versa.

Only that research hasn't panned out. It turns out there is no genetic explanation for this striking statistical correlation. In fact, in one study of more than eight thousand individuals, those with autoimmune disorders had an increased risk of depression, independent of their genetic risk for depression. And those with depression were more likely to develop autoimmune disorders, independent of their genetic risk for autoimmune disease.

Seeing the brain's and body's immune systems as intricately interconnected—with microglia at the center of the story—helps us to understand at last why this is true in families like the Harrisons.

And yet, Katie tells me, her family has never viewed their emotional and mental health struggles as related to their physical and autoimmune issues. "The extent of our family's understanding of the whole brain-body link is maybe some recognition that having connective tissue disease must make life more difficult for my mother, so of course she would be depressed," she says.

Katie and her family are hardly alone. Most individuals with brain-related diseases do not view their conditions as having anything to do with the state of their brain's immune cells because they don't even know that these cells exist. And because they don't understand that their brain's immune system can be triggered into disease, just as their body's immune system can be, they don't consider new approaches that might help them find better brain health, and a better life.

The next time I speak with Katie over Skype she confides that "all of this information is anxiety-making for me." I'm not surprised. This information is a lot to take in—for me too. And yet this science also gives us a promising new window into better understanding an array of intractable disorders of the mind so that we can pursue novel ways to treat and alleviate suffering for so many.

Recently, Katie tells me, her integrative physician informed her that she has "high" inflammatory biomarkers for C-reactive protein. "And yet," she says, "no doctor has ever mentioned to me that higher inflammatory markers are associated with more mental health disorders." Clearly, she says, her family tree is a case study for this feedback loop between the body's and brain's immune functions. "Yet in all these years, I've never heard any practitioner even utter the words 'microglia' or 'neuroinflammation' or even talk about brain circuitry!"

If we could see the brain in this new light, Katie opines, "researchers might be focusing more on microglia, and neuroinflammation, and families like mine might be benefiting from new avenues of treatment. After all, medical science works to help our physical immune systems and physical health to be better. I would like to have a better immune system, and healthier microglia, in my brain too."

This lag time between where science is and where patients are is hardly surprising. As the scientific philosopher Thomas Kuhn said (to paraphrase loosely), it takes about twenty years for a new paradigm shift in the research world to reach the doctor's office.

That said, I reassure Katie that even as psychiatry continues to turn a blind eye to this science, a lot is happening in research labs to identify new avenues of understanding that can lead to treatment.

But in order to more clearly unveil how science can fully utilize microglia's good side, scientists have also had to figure out how microglia interact with the body's physical immune system in the first place. If physical diseases and inflammation in the body, like lupus, gum disease, bacterial infections, Crohn's disease—or simply having a higher inflammatory response from chronic stress, without any signs of physical disease—are also triggering inflammation in the brain, and if this in turn prompts microglia to overenthusiastically destroy synapses, exactly how is the body's immune system talking directly to the brain?

How in the world *are* the white blood cells of the body and the brain's microglial cells sending messages back and forth?

FIVE

A Bridge to the Brain

I T HAS SOMETIMES BEEN SAID THAT WE KNOW MORE ABOUT THE moons of Jupiter and the rings of Saturn than about what is inside our skulls. From the time he was a young grad student, Jonathan Kipnis—who goes by Jony (pronounced Yo-ny) to friends—wanted to change that. As early as 2003—almost a decade before the brain-immune revolution in science—Kipnis, who was at the time four years into pursuing his doctorate, was convinced that the body's immune system had to be playing a role in psychiatric disorders and neurological autoimmune diseases.

Today, Kipnis, who is in his forties, with close-trimmed dark brown hair newly flecked with gray and a short, scruffy beard that gives him a hip, youthful appearance you might associate more with a philosophy professor than a scientist, serves as director of the Center for Brain Immunology and Glia (or BIG) and chairman of the department of neuroscience at the University of Virginia. But back in 2003, when he was wrapping up his Ph.D. at the Weizmann Institute of Science in Rehovot, Israel, many of Kipnis's professors were dubious about their young student.

In one of his earliest experiments as a graduate student, Kipnis had experimented with altering the physical immune system in mice by removing their T-cells. T-cells are the army sergeants of your immune system; they tell the immune system's troops—the white blood cells—when and where to mount attacks against infections or pathogens infiltrating your body. When Kipnis removed T-cells, he found something surprising: It dramatically altered brain function. Suddenly, these mice didn't learn as well as normal mice did. Conversely, when he reintroduced T-cells, the mice learned normally again.

It seemed to young Kipnis that this was a crucial area of inquiry that no one was exploring. But when he showed his first paper demonstrating a link between T-cells and mouse cognition to his colleagues, "it was extremely poorly accepted by everyone I presented it to," he says. "My professors insisted I had to be wrong; it couldn't be true." One colleague went so far as to tell Kipnis, with no small degree of sarcasm, "In the future, if I have to invite someone to speak on an esoteric subject at a conference, I would invite you."

Kipnis remained undaunted. By the time he graduated with his Ph.D. in 2004, he'd coauthored seven papers on what he saw as a likely, albeit mysterious, brain-immune link. In one prescient study he questioned whether an injury to the central nervous system could cause the mysterious housekeeper cells in the brain, microglia, to somehow contribute to neuronal degeneration. He wondered whether the body's T-cells might somehow even be in dialogue with microglia in ways that might influence neurological autoimmune diseases and psychiatric conditions.

"Although fifteen years ago there was still thought to be no known link between the two systems, I also knew this couldn't be true; the immune system had to be somehow impacting the brain," Kipnis says, in his impassioned Russian-Israeli accent (he grew up in the Republic of Georgia before moving to Israel as a teenager in 1990, when his family fled the collapsing Soviet Union). He points to what medicine has long referred to as "sickness behavior": When individuals are depressed, "they also feel physically sick. They lose their appetite. They are so fatigued they can't move." He felt there had to be a brain-body link.

In 2005, after finishing his Ph.D. and a short postdoc stint at the Weizmann Institute, Kipnis headed to the United States. In 2010, he published work showing that when the body's T-cells expressed certain chemicals, this also led to cognitive impairment in mice. Clearly, T-cells were exerting some kind of direct influence over the brain.

By then, interest was also heating up around microglial cells and their newly understood role—as immune cells—in governing brain health. Perhaps the body's immune system and the brain's immune cells did chat? But how in the world was an overactive immune response in the body talking to and influencing the behavior of microglia? By 2015, says Kipnis, "Ask any neurologist and they would be the first to tell you that it was becoming clear to them that neurological disorders are always associated with some degree of immune system dysfunction."

Yet even as researchers were categorically beginning to accept that this link existed, "we still couldn't study the brain-immune interaction on a *mechanistic* level," explains Kipnis. "We knew some really major piece was missing in our understanding."

It was an enormous scientific gap.

The Scientists Who Thought to Look

Many great scientists seem to have a knack for hiring and mentoring great postdocs, and Jony Kipnis is certainly no exception.

In 2015, Kipnis's postdoc Antoine Louveau, Ph.D., who was working in his lab at the BIG center at UVA, had certainly been taught, as all scientists had, that the brain was the only major organ that lacked a direct physical connection to the immune system.

Still, more and more evidence was countering that dogma. For instance, other researchers had recently shown that when they injected T-cells into the brain (in animal studies) they somehow eventually made their way down into the body and showed up in the cervical lymph nodes.

This really didn't make sense. If the body's and brain's immune systems had no anatomical bridge connecting them, how in the world were T-cells injected into the brain showing up in other parts of the body?

"Not every cell that gets injected into the brain gets into the body, but some cells do get in," says Kipnis. "So the question was, how do they get into the body? And how do they leave the brain?"

Kipnis and Louveau had become deeply intrigued by an area just outside of the brain known as the meningeal spaces. The meningeal spaces are made up of layers of membranes that wrap around the brain just beneath the skull, like a thin, protective cap. At the time, it was thought that the primary function of these meningeal membranes was to carry the cerebrospinal fluid (CSF) that buoys the brain.

Kipnis and Louveau wanted to look at these meningeal spaces more closely. Louveau found a way to firmly affix the meninges of mouse brains to the skull cap before peeling off the meningeal membrane. Then he examined it, in its natural setting, so that he could see this vast network while it was still completely intact (usually researchers removed the tissue first, and manipulated it onto a slide) and *then* dissect it.

No one had ever done it this way before.

What Louveau saw, peering through his microscope at brain tissue, shocked him. He was staring at something that shouldn't be on his slide: There appeared to be lymphatic vessels hidden inside the meningeal spaces.

He realized immediately the magnitude of what he was looking at.

The lymphatic system, which is part of the body's circulatory system, is responsible for ferrying immune cells—the T-cells' army of white blood cells—throughout the body. These lymphatic vessels course through your body the way the earth's underground springs run through and beneath the land.

For instance, say you are jogging on a dirt path and you trip and skin your knee. T-cells send their army of white blood cells marching into the tissue around your knee in order to protect your body from the bacteria, fungi, and microbes in the dirt and on the stones upon which you fell. This immune brigade circulates to the precise site of your wound through this intricate, waterway-like system of lymphatic vessels.

For hundreds of years, medical textbooks had taught that, anatomically, these lymphatic vessels could not, and did not, exist in your brain.

The fact that lymphatic vessels were not found in the brain was considered part of the proof that your circulating immune system had no jurisdiction over your mind.

But Louveau was staring down at an array of lymphatic vessels inside the meningeal membranes surrounding the brain—exactly where they weren't supposed to be.

"I called Jony to come take a look through my microscope," Louveau recalls. Then he calmly told Kipnis, "I think we have something here."

Jony knew immediately what this finding meant. But he also knew it could be a mistake. Never one to finesse his answers, he told Louveau, "Let's make sure we're right."

"I was initially skeptical," Kipnis recalls. "I really did not believe there were structures in the human body that we were not aware of. I thought the body was mapped, and the big discoveries ended somewhere around the middle of the last century."

Kipnis immediately "went to a colleague and asked, do you have a marker that we could use to mark these vessels and make sure they are immune vessels?" (Scientists use fluorescent markers to make protein molecules that are unique to a specific organ or system—in this case, lymphatic vessels—glow with bright fluorescence when illuminated by a particular wavelength of light.)

Kipnis's colleague had what he needed. But, he told Kipnis, "You're wasting my markers."

Kipnis and Louveau added the visual markers to the tissue. And there it was again: Lymphatic vessels inside the skull caps of the mouse brains were indeed glowing in bright fluorescent definition. They were irrefutably there. "It was a very powerful moment," Kipnis says, in something of an understatement.

The presence of these previously hidden immune vessels raised the question: Was it possible that these vessels were somehow involved in ferrying signals from the brain to the immune system?

First, Kipnis and Louveau needed to replicate their experiments. They spent the next six months collaborating with experts who studied lymphatic vessels. Each time they repeated their study, they got the same results. "We had to completely prove to *ourselves* that we were right before we published," Kipnis says.

When Kipnis showed colleagues in the department what they had, "they told us we would need to rewrite the textbooks," recalls Kipnis. After two decades of looking for a bridge between the brain and the immune system, Kipnis's lab had found the missing link.

In 2015 they published their discovery. It stunned the scientific world. In their paper's conclusion, Kipnis and his colleagues summed up by saying that "current dogmas regarding . . . the immune privilege of the brain should be revisited."

Still, doubters persisted: Perhaps this was the case in the mouse brain, but not in the human brain? The next year, Kipnis's lab, in collaboration with a group of investigators at NIH, was able to prove that these lymphatic vessels were present not just in mouse brains, but also in human brains.

The group of investigators studied five healthy volunteers—two men and three women—by injecting a safe contrast dye into their brain and then scanning the brains in an MRI machine. Afterward, they enlarged the 3-D images, which allowed them to see the same lymphatic vessels they'd seen in mice. Kipnis and his colleagues began to create the first map of the meningeal lymphatic networks within the human brain, giving scientists an entirely new avenue into better understanding and treating human neurological and immune disorders.

In a single experiment in Kipnis's lab, the idea that the brain didn't physically interconnect with the body's immune system had been overturned. Other researchers around the world have since replicated and corroborated their findings. In 2015, Kipnis's finding was called one of the ten most important scientific breakthroughs of the year by *Science* magazine.

But, says Kipnis, "I certainly didn't feel we deserved this recognition. I felt we had to figure out *how* this finding mattered in helping us better understand not just the workings of the brain but the diseases that plague it."

Pipeline to the Brain

The discovery of these lymphatic vessels bridging between brain and body opened up myriad questions for researchers in terms of disease relevance. For instance, in the body, the lymphatic system not only ferries immune cells to the site of infection to fight off a foreign intruder, but also carries away the resulting cellular debris from that area and disposes of it. This is a well-choreographed set of immensely important protective steps.

But sometimes, in autoimmune diseases such as rheumatoid arthritis, lupus, and MS (or in diseases like mine, Guillain-Barré), the immune system becomes overactive and sends the wrong message. Immune cells attack healthy tissue and cause more harm.

Now, says Kipnis, "We can begin to ask mechanistic questions about this process in the brain. Now we *know* the brain is like every other tissue, connected to the peripheral immune system through meningeal lymphatic vessels."

Could it be that the lymphatic system was somehow triggering immune cells in the brain—microglia—to carry out an overactive immune response, creating neuroinflammation, or triggering microglia to eat away at synapses?

At the moment, we don't know. What we do know, says Kipnis, is that the meningeal spaces house immune cells from the body that can release cytokines, which in turn influence brain circuitry.

Kipnis's work has raised another line of inquiry. Could it be—given that these lymphatic vessels are supposed to help to clean out the brain—that in some cases the brain wasn't being properly cleaned?

Kipnis goes on, excitedly, "We believe that for every neurological disease that has an immune component to it, these vessels may play a major role." Take Alzheimer's disease, for instance. "We know that in Alzheimer's there are accumulations of big protein chunks in the brain. We think they could be accumulating because they're not being efficiently removed by these vessels."

We need, he says, "the sewage system to be working." And we know, in Alzheimer's, that "with aging, these lymph vessels become reduced in size. In part, it's a plumbing issue. It's possible that the health of these vessels helps determine when Alzheimer's starts. So, what if we can prevent or unplug the clogging of these vessels?"

Or, says Kipnis, smiling, "What if we could push back the age at which people develop Alzheimer's to a much later age, say a hundred and sixty or so!" Kipnis's own grandmother is ninety-three years old, he tells me, "and signs of Alzheimer's are just starting. But if we pushed the age at which Alzheimer's starts in the brain back further it could become a very different disease, one we don't even worry about getting."

Kipnis uses this analogy. Think of this lymphatic system, as it clears debris, as being like "the trash collection in your home," he says. "If I look through your trash for a long time, I will know what foods or chemicals are in your body. The same thing is happening with the immune system. The immune system's job is to be closely observing all the time, because if something is wrong, its job is to fix it." So, he goes on, "If the immune system sees bacteria or other invaders, they want to rush in and take care of it. But if, at the site of the infection, the body sends the wrong message to the immune cells, or the immune cells don't properly understand the message, and they send out the wrong response, we have a problem."

Now just imagine, he continues, "if we can find a way to intervene in faulty messages coming to or from the brain to the lymphatic vessel system. What if we can intercept the wrong message, and maybe deliver the right message to the immune system instead?"

Together, Jonathan Kipnis and Beth Stevens and their teams have shown us two extraordinary things. Not only is the brain an intricately sensitive immune organ, full of tiny and sometimes hyperactive immune cells that no one previously knew affected synaptic health, but the brain is also physically connected to, and engaged in, a constant dialogue with the body's immune system.

T-cells, white blood cells, and—in some way we don't yet understand—

microglia are cross-chatting through these lymphatic vessels that travel up from the body's lymphatic vessels into the meningeal spaces—like open pipelines—into the brain.*

To put it simply, when the body is ill, white blood cells signal inflammatory molecules to alert microglia, *Hey! We've got a problem here! You'd better go on offense!* Inflammatory chemical messengers signal glia to ramp up their aggressive action, exerting what researchers call "direct toxicity" on the brain.

The work of this small band of researchers and the discovery of microglia's powers have given science an entirely new unifying theory of brain-related disease:

A (Microglia are the white blood cells of the brain.)
+
B (The body's immune cells directly communicate with the brain through tiny vessels that bridge body and brain.)
= C (Anything that triggers disease in the body can easily influence the immune system in the brain and trigger disease there too.)

There are still many questions to answer, of course, about the interplay between microglia and the immune signals that traverse the meningeal immune vessel pathways into the brain.†

I pose one such question to Kipnis: "If we know that the brain's immune messages are coming through the lymphatic vessels to the body

---

* The brain and body may also engage in back-and-forth messaging through other portals. Microglia appear to be key players in maintaining the blood-brain barrier. When microglia sense damage to the barrier, they rush out to help clean up the debris from any dead or damaged cells and seal off the site of the injury. Are microglia also getting messages from the immune system in this way? We don't know.

† The brain's meningeal lymphatic vessels, Kipnis points out, shouldn't be confused with the body's glymphatic system. The glymphatic system, which surrounds the spine and skull and carries the body's cerebrospinal fluid through brain tissue, thus "washing" the brain clear of debris, says Kipnis, "is directly controlled by the meningeal lymphatic vessels."

and the immune system, and if we also know that brain-based disorders from depression to Alzheimer's are disorders in which microglia either spew out inflammatory markers that damage neurons and engulf and destroy synapses, or fail to take out the right garbage, then how do we intervene and keep microglia from sending and receiving the wrong messages in the first place?"

Kipnis tells me, "This is the two-hundred-million-dollar question. But now that we are including the immune system in the question of how circuits are affected, everything will become more clear to us."

Meanwhile, what Kipnis's discovery tells us, irrefutably, is that the body's immune system has direct access to the brain.

And what Beth Stevens's discoveries indicate is that when microglia get "bad news" from the body, they begin to erroneously destroy neural circuitry.

This is what I call the microglial universal theory of disease.

SIX

"It Seems There Are No New Solutions"

I T'S A BREEZY SATURDAY IN LATE AUGUST IN COS COB, CONNECTI-
cut. It's one of those mornings when the wind coming in off the water
sings through the sailboat halyards—a familiar sound in the seaside
town.

It's sunny, not too humid, the kind of summer day you hope for.
Heather Somers's nineteen-year-old twins—her daughter and son—are
about to head back to college for their sophomore year. Heather knows
she should be finding a way to create some family come-together time—
make blueberry pancakes, organize a day on the boat, offer a little moth-
erly calm and reassurance after what's been a difficult few days—in fact,
to be honest, a hell of a summer. Just the way she's always managed to
do, despite whatever crises she's been helping her daughter, son, or husband—or herself—through.

Heather is the queen of working the problem, hand-holding, strategiz-
ing, fixing seemingly unsolvable family crises. But today she is not making Mom magic happen. Today she has had it. She is, she thinks, like the
Giving Tree in the children's story she used to read to the twins when
they were very young: She has given so much of her life energy to those
she loves that today, she has nothing left to give.

And so, Heather, who is fifty-five and a high school teacher, is hiding from her family inside a dilapidated wooden tree house in the backyard, her back resting against the splintering planks that she once, when the children were small, painted robin's-egg blue. She has situated herself where no one can see or hear her from the house.

She is letting "it"—something "dark and twisty" she can't name or explain—rise up in her and come out in unpretty sobs. She cries until saliva and snot begin to spread in a wet stain across her blue T-shirt. At some point, she lightly dozes from the sheer exhaustion of crying.

The sound of voices wakes her from her semifugue state.

Heather's husband and daughter and son are calling for her as they walk through the yard. "Mom!" her daughter calls. "Mom!" her son calls. "Heather!" her husband calls.

Heather doesn't answer, not because she's trying to be secretive or sly, but because she does not want the twins to see her like this. They *must* not see her like this, she thinks.

That's when she hears her daughter, Jane, clambering up the ladder. Jane sticks her head in the small door of the tree house. "Mom?" she says. "Mom, what are you doing up here? Are you *okay*? Where have you *been*? We've been looking for you *everywhere*!" And then, taking in her mom's face, "Oh my *God*, Mom, what *happened*?"

A few weeks later, Heather is sitting at my kitchen table, sipping a green smoothie she's brought with her; I'm nursing my usual mug of Earl Grey. We've connected via mutual friends whom she's visiting near me while dropping Jane off at college.

That terrible morning at home, Heather tells me, had started out okay enough. She'd done her yoga practice, eaten two slices of paleo toast. She'd fed the dog and the cat, organized the fridge, then gone out to sweep the patio.

"The twins had dumped their stuff everywhere; there were piles of dirty clothes, shoes, and stuff they'd stacked up so they'd remember to take it back to school in a few weeks," Heather says. "They'd left open bags of snacks spilling across the countertops from the night before." Heather began straightening up, she says, "I think to distract myself."

Heather was tired; the day before, Jane had called her in a panic. Jane had been finishing up the last two weeks of an internship in New York City, and she had called to say she wasn't feeling well. She sounded distraught—unable to catch her breath, talking and crying almost incoherently. Heather told Jane that she would come in to the city and get her. It was only an hour away.

Dropping everything to care for her daughter was hardly new for Heather. "Even back in high school, Jane would have terrible panic attacks," Heather says. "She went on Prozac, and we did everything we could to help her. When she was really struggling, and her anxiety was unmanageable, I'd walk with her around her high school's athletic fields and help her do deep breathing," Heather recalls. Jane also had a strong teenage temper, "and when she wasn't needing me to comfort her, she was dumping on me, sometimes in the same breath."

Heather jumped on an express train to New York to bring Jane back home. "It was the third crisis with Jane this summer," Heather says. "I knew she needed to be at home where she could see her therapist and psychiatrist."

"Jane was so worried about what would happen if she didn't complete her internship, she was in a panic the whole way home on the train," Heather says. "I tried to say the right things, but no matter what I said— whether I validated her feelings, or encouraged her that she would get through this—she was furious with me either way. And even though I know she gets upset with me because I'm the only person she feels safe enough to vent her frustrations on, it pushes my buttons. I have to take a deep breath and let go of how I feel, because she's not feeling well. And I am, after all, the grown-up."

That morning, as Heather swept the patio and her kids slept in, she looked up at the tree house that her husband, David, had built when the kids were little. "It was cool for an August morning," she says. "There was something in the breeze, maybe the smell of fall coming, that made me think back to the mornings that I'd taken the kids some snacks in the tree house, or how they'd spent so many afternoons in there pretending to be pirates or reading piles of library books. I still remember serving Jane and her friends tea and cinnamon toast on a red-checked tablecloth." And then, Heather says, "I just felt tears flooding down my face."

Heather suffers from her own health crises too. She has rheumatoid arthritis, an autoimmune disease in which the body's immune system mistakenly attacks the body's joints, causing pain and inflammation. She has been managing her symptoms for fifteen years. "It's mostly my hands and shoulders that just completely freeze up," she says. Recently, she was also diagnosed with Sjogren's syndrome, an autoimmune disease in which the immune system attacks the glands that produce saliva, leading to dry eyes and dry mouth as well as osteoarthritis, a degeneration of joint cartilage and the underlying bone.

In addition to helping to manage Jane's panic attacks, Heather worries about her husband, David, too. Ten years ago, David, who was at the time an active duty Army physician, was riding in a jeep that drove over an improvised explosive device, or IED, in Afghanistan and sent him flying through the windshield. He'd suffered a traumatic brain injury, or TBI. He walked well by now, with a cane, but he still suffered from post-concussion syndrome.

"When he was deployed, the kids slept with me," Heather recalls. "That went on for two years. I'd be so scared about what would happen to him that at times I thought I was losing my mind. And then, after his accident, when he came home, I'd wake him up at night, just to tell him that I was too scared about what the future might bring to sleep."

Like many veterans, David received counseling and medication from the Veterans Administration, but, like many veterans' wives, Heather found herself making up the considerable gap, taking care of every chore that needed doing at home, and most of the parenting duties, as David— understandably—focused on simply learning to walk and drive again. "He's a wonderful father and husband," Heather underscores. "So patient, so compassionate, very smart. But it has been a lot, and at some point we worry that he'll have to live in the veterans' home, because depending on how he is as we get older, and where my RA is as I get older, I might not be able to take care of him when we're seventy."

In the face of all this, Heather's own anxiety levels have increased, year by difficult year. "I had low-grade anxiety even back in high school," she tells me. "I can remember looking up the words 'nervous breakdown' in the dictionary in the high school library when I was probably fourteen." Still, the feelings weren't unmanageable, and Heather was able to navigate her way through them.

Once, in her twenties, after Heather had moved to New York City to start her first job, she recalls "going to a party and walking in in a fuchsia dress, and realizing I'd dressed completely wrong. Everyone else was in black." While this is a garden-variety type of anxiety—showing up wearing the wrong thing and worrying that you stand out—for Heather, it engendered, she says, "utter anxiety and self-loathing. I was very insecure and unsure of myself. I felt intimidated by everyone." During those years Heather noticed one thing, she says, that she never forgot: "I could see that other people my age were more comfortable in their own skin. I was different that way. I was not comfortable in my own skin."

For years she gritted it out, trying to ignore the "white noise" of anxiety increasingly churning inside her, always playing like background music. And then, in her midthirties—fifteen years ago, even before Dave's accident—she received the diagnosis of dysthymia, a form of depression marked by a loss of interest in life, as well as generalized anxiety disorder (GAD). A year later, she was diagnosed with RA. "I knew something had to change," she says. So Heather developed a serious yoga practice that has helped calm her body and mind, and she has even become a mindfulness-based stress reduction teacher. She also developed programming for a mindfulness wellness course for students at her school. "If it weren't for yoga and mindfulness and eating green, I probably wouldn't have made it this far, given everything we've faced."

She verbally ticks off, without a trace of self-pity, a list of the meds she's on: NSAIDs, Pepcid, Plaquenil for RA, trazodone for insomnia, Zoloft for depression and anxiety. "There have been so many crises with my health, and my husband's, and now my daughter, that I have come to live expecting an emergency," she says. "Even our son, Ian, had a lot of trouble getting through high school. He has a diagnosed learning disability, dysgraphia, and ADHD, and he's had migraines for years. He's also a remarkable pianist." She smiles. It has, she admits, "been a lot to care-manage."

Some of their constellation of problems, she thinks, is "just family genetics, some of it is all the shit we've been dealt, and the trauma and high-octane stress of it all. It's a genetic and environmental double whammy."

Just talking about all this makes Heather feel "a little sick," she tells me, as she sits in my kitchen. She runs her hands up and down her arms as if she's cold, though it's an 80-degree day. "My arms tingle all the time. I'm too stressed to eat, I can't bring myself to make dinner," she says. "To be honest, I just want to go lie down and binge Netflix. I can't deal. I feel like my memory is shot too. I can't remember anything. At times it's more than I can do to just hold a conversation." Sometimes, she says, "I get this gut-wrenching sensation, as if I'm being punched in the stomach, and I have to run to the bathroom. I have this terrible sense that nothing will ever be okay again."

Nevertheless, she says, "I've always just worked the problem that's in front of me no matter how I feel. For years and years I've secretly thought, *When am I going to fall apart?* I guess the answer is now."

Which brings us back to the morning of the tree house incident.

That morning, sweeping the patio, says Heather, "It was all too much." Her whole body hurt. Hell, her whole brain hurt. "I was so overcome with all the crushing anxiety that comes with having autoimmune issues, and all the icky worry and confusion that comes with sorting out my daughter's issues—being a caregiver ad infinitum—that I just hid." Heather sips her kale-carrot-ginger concoction. "In the tree house. For hours. Several of which my family spent looking for me. And here I am, *a grown-ass fifty-five-year-old woman.*"

During her hours in the tree house, it went through her mind, not for the first time, she says, that "women work hard and take care of everyone and we pretend to be okay, even when we are not okay. We just shove aside our needs for so long in order to raise these beautiful kids." She finger-combs her pixie-short dark hair away from her temples as her words spill out. "And the idea, all along, is that everyone *will be okay.* We'll have a happy family, kids sailing off to college, healthy and well prepared, with Mom supporting them one hundred percent. But then suddenly, they are not okay. And I am not okay. And there is a kind of primal wound around that. Now my daughter is suffering so much, and despite all my best efforts *I can't help her.* She's not always nice to me, she pulls me in while also lashing out at me. My family takes their stuff out on me. All the output of caregiving for decades, while I hid my fear and my fatigue and my pain to assuage theirs—I did all that because I wanted *them* to be okay. And yet we all are still suffering anyway. *And it wasn't supposed to be this way.*"

And that morning, it all hit her, and the sobs came out, in the musty old wooden tree house.

"We are a family that tells people a story," Heather explains. "We say I have RA, and that our son has ADD. We don't tell anyone about Dave's struggles, or my taking Zoloft, or Jane's being on Prozac and having to leave school for a few weeks during her freshman year because of her panic attacks.

"We live with all of this in the era of the Facebook Effect, in a social cocoon of people who portray their perfect lives and vacations and awards. We don't post about what's going on in our house. People wouldn't be open to hearing the truth about that."

Recently, for example, when Heather took the train to New York to help Jane, she told a close friend, who knows that Jane has a panic disorder and is on medication and in therapy. And her friend said, "Oh my God, what's the drama today?"

"We are all suffering," Heather went on, "but Jane is suffering the most, bearing the brunt of familial genetics and the trauma of growing up with the fact that her dad's life was almost lost, and witnessing two parents with chronic illnesses."

Heather has seen in her own family what happens as people with brain-related issues get older. Her grandfather died, very early, from Alzheimer's, and two of her father's three brothers suffered from severe depression. "I saw how each of their mental health unraveled, and it wasn't good." It was, she says, like watching all the stitches being slowly pulled out from an heirloom patch quilt, one by one.

"I don't want our lives to go that way," Heather says. "It seems to me that brain-related disorders are the big black hole of health care, and that means patients like my family are stuck inside that big black hole too.

"It seems like there are no new solutions, we are stuck with the same-old, same-old: diet, exercise, meds, cognitive behavioral therapy, dialectical behavioral therapy, etcetera etcetera," she says. "It's just *not enough*. I have to help myself, and my family, get past this place where we are just surviving all the shit we have."

It is cold comfort, Heather says, but she knows that in some ways, she and her family are not alone—especially not in terms of what they face with their daughter. She pulls a recent *New York Times* article out of her bright green leather bag and places it on my kitchen table. It's an article I've already read; it discusses today's adolescent anxiety crisis. Many of my parent and educator friends have sent it around to each other.

Heather switches from her brown tortoiseshell glasses to a pair of red reading glasses and looks down at the article. She jabs her manicured index finger on a paragraph before passing it to me. The section she's underlined talks about how the rising number of adolescents facing anxiety, depression, learning issues, and mood disorders is affecting schools like hers. "At every educational conference I go to, teachers are saying, 'Look, there is something big and mysterious going on with our students, and especially our girls, our girls are *not* okay.'"

The trend is, I agree with Heather, undeniable.

A Growing Trend of Disturbing Proportions

Statistics on adolescent female depression and anxiety in the United States today are staggering. In one recent year, one in six adolescent girls reported an episode of clinical depression. In a study of one hundred thousand children between 2009 and 2014, researchers found that depression in girls is occurring earlier, often by age eleven. By the time teens hit the age of seventeen, a shocking 36 percent of girls report having had a depressive episode—often marked by a sense of "worthlessness, shame, guilt, and insomnia."

These aren't minor episodes of not feeling great, either. According to the National Institutes of Health, in 2016, three million adolescents between twelve and seventeen faced one or more episodes of major depressive disorder.

That's one out of every seven teenagers in the United States—again, mostly girls.

This epidemic is affecting boys too, of course. Boys suffer from depression and anxiety at about one-third the rate of girls, but boys face more learning disabilities, autism spectrum disorders, behavioral disorders, and attention deficit disorders (which often coexist with anxiety).

To date, the jury is out on what's fueling this trend. There is a litany of the usual suspects: Has today's teenage generation—often referred to as iGen, which includes Heather's twins, Jane and Ian—been adversely impacted by a toxic culture of social media? Have they been unduly stressed by today's high-stakes, stress-mill, get-into-college educational system? Or did their parents coddle them too much, so that when they meet up with real life stressors they don't have the coping skills they need, and are too easily triggered when they face small obstacles, or situations that they can't control? And what part of this trend is due to a greater diagnosis of these disorders?

It's a head-scratching unsolved mystery in social science.

Helping Jane make her way through adolescence has sometimes been so painful for Heather that, she posits, "I think it worsened my own anxiety and medical issues." Indeed, it was after Jane developed her psychiatric disorders that Heather developed her second autoimmune disease—Sjogren's. "Jane's stress is my stress," she says.

Even Heather's recent drop-everything trip to bring Jane home for an emergency visit with her psychiatrist when Jane had her full-blown panic attack in New York City is, according to recent studies, hardly unique to Heather's family. Adolescent psychiatry practices and private residential treatment centers for adolescents—what's now known as the troubled teen industry—are doing a booming business.

In fact, 535 pediatricians from small-town doctors' offices as well as big-name hospitals reported in 2014 that the rise in rates of kids between the ages of six and seventeen exhibiting signs of anxiety and depressive disorders was unprecedented. Between 2010 and 2013 the rate of children diagnosed with anxiety disorders rose by a staggering 72 percent, depression by 47 percent, autism by 52 percent, and eating disorders by 29 percent.

In 2017, two Vanderbilt pediatricians examined trends at thirty-two pediatric hospitals across the country and found that the percentage of children and teens being hospitalized for suicidal thoughts or actions in the United States had doubled in the past decade. The study's authors write, "We'd noticed over the past two or three years that an increasing number of our hospital beds were not being used for kids with pneumonia . . . they were being used for kids awaiting placement because they were suicidal."

It's not just teens and children who are suffering, of course; adults, like Heather, who are in midlife are also facing surprising increases in mental health and cognitive disorders.

Rates of adult suicide have surged over the past decade in the United States — in 2018 the suicide rate reached a fifty-year peak. Not surprisingly, as depression rates rise, so do rates of addiction. In 2017 alone, fifty-two thousand Americans died of heroin overdoses.

Alzheimer's is, of course, its own public health crisis. Today, five million, or one in nine, adults over sixty-five suffer from Alzheimer's disease. And while the aging of Americans plays a role in Alzheimer's, it doesn't account for the fact that more Americans are developing the disease at increasingly younger ages, in what's known as early-onset Alzheimer's.

What's driving rising rates of early-onset Alzheimer's, midlife distress, depression, addiction, and suicide, and an epidemic of debilitating anxiety in adolescents?

And what do tiny little microglial cells have to do with all this?

The answers are as paradigm-shifting as they are surprising. And they offer us a new way of thinking about the brain, one that promises to lead families like Heather's toward new avenues of relief and recovery.

SEVEN

A Modern Braindemic

D ORI SCHAFER—THE "SCIENTIFIC DAUGHTER" OF BETH STEVENS and "scientific granddaughter" of Ben Barres—is a dedicated, down-to-earth researcher who, ten years after bursting onto the scientific scene as a young investigator in Stevens's lab and pioneering the imaging technique that demonstrates how microglia destroy synapses, is now an award-winning scientist in her own right. Schafer serves as a professor of neurobiology at the University of Massachusetts Medical School, where she continues to study the role of microglia in an array of diseases across the life span.*

I turn to Schafer—the first scientist ever to observe microglia destroying synapses in real time under her microscope—and ask whether she thinks the emerging science on microglia gives us new insight into the increasing number of neuropsychiatric, neurodevelopmental, and neurodegenerative disorders we're seeing in so many different age groups.

Schafer agrees it's a crucial question for our time, and one that scientists are only now beginning to ask. "We're just scratching the surface as

---

* Dori Schafer, Ph.D., received the 2017–2019 Young Investigator Grant from the Brain & Behavior Research Foundation, a nonprofit that funds mental health research, as well as the 2016–2018 Child Health Research Award from the Charles H. Hood Foundation.

to the environmental effects of our modern environment on brain health," she admits. "And once we ask that question, we have to also ask how today's environment is affecting microglia" in specific. Is something in our environment triggering these tiny cells to express more inflammatory factors, and potentially eat more synapses—resulting in more disease?

"Microglia have always been around, even if we are just starting to understand their role in brain health," she explains. "So *that* hasn't changed."

Genetics, of course, plays a role in which individuals are more prone to develop a certain disease at a certain time in their lives. Whatever genes made it more likely that Heather would grow up to develop dysthymia made it more likely that her daughter, Jane, would face anxiety too.

But it is also true that there are no genetic epidemics. Genes simply don't change that fast, over the course of a generation. Genetics alone cannot account for these growing trends.

To some degree, Schafer points out, higher diagnosis rates are also at play—more primary care providers are trained to catch psychiatric disorders and addiction in patients and intervene as early as possible, and more pediatricians are on alert for neurocognitive and neuropsychiatric disorders in children and teens. And, as more Americans are aging, public health efforts have educated us about Alzheimer's—so more aging Americans are getting diagnosed sooner.

But better diagnosis rates can't account for these staggering trends.

So, if microglia haven't changed, and there are no genetic epidemics, and better diagnosis rates alone don't account for these rising rates, what does?

"Our environment has changed a great deal in such a very short period of time," says Schafer, who has a wide, open smile and shoulder-length ash-blond hair. "We've undergone enormous changes in our diet in the past hundred years, we've been exposed to more toxic chemicals in the environment—and there are just so many more chronic societal stressors in everyday life." For instance, she says, "We know that kids growing up, especially girls, are exposed to a lot of psychological pressures in our culture that are novel."

For instance, Schafer points out, "Growing up as young women, we receive constant cues regarding body image and gender roles. We are also continuously on guard, witnessing sexual harassment at school and in the workplace and seeing how women are depicted in the media." Girls can't help but worry, at every turn, whether they measure up—and whether they are safe in our society.

Given the ubiquitous nature of social media in today's digital age, most girls are exposed to these chronic psychosocial stressors—a never-ending stream of stories about incidents of violence or inequities perpetrated against women, as well as commentaries critiquing images of the female body—on a relatively constant basis. And, often, girls manage any stress they may feel alone—in today's society, we frequently lack the kind of wider community connectedness and extended-family closeness that can help buffer the effects of stress.

"And we know that chronic stress changes the body and brain," adds Schafer.

All of these toxic exposures, Schafer says, "result in peripheral immune events that affect the brain and vice versa. The interesting thing is that when these events affect the brain, they affect microglia too."

Part of the problem in elucidating this trend is that "it's hard to nail down any single environmental event," she says. Instead, myriad triggers—exposure to environmental chemicals, unhealthy diets, chronic stressors—add up, affecting the brain's microglial immune response in a cumulative way, which could lead to more runaway inflammation and more synapse loss over time.

We also know that microglia communicate with the body's immune cells. And we know that microglia can be easily triggered to spark neuroinflammation, or chip away at brain synapses, by the very same stimuli that elicit inflammation in the body.

So why have microglia, which should function as the protective immune cells of the brain, morphed so dramatically in so many individuals from behaving as the angels of the brain into dangerous assassin cells in the course of just a few generations? Well, that requires a brief trip back in time.

Microbes, Pathogens, and Human Behavior

Let's pretend that it's five hundred years ago, circa 1500, and you live in a small feudal village somewhere in Europe, where infectious diseases such as whooping cough, measles, and tuberculosis (which once had a 50 percent mortality rate) regularly make their way through your town. Many children do not survive to adulthood.

If tuberculosis starts to wend its way through your village, and you are unlucky enough to catch it, your immune system will immediately crank up inflammatory infection-fighting immune cells to do battle against the pathogen. Levels of inflammatory cytokines in your body will skyrocket.

It probably won't surprise you at this point to learn that, while you are ill, your microglial cells are also mounting an inflammatory response in your brain.

And here's where it gets really interesting. Evolutionary biologists believe that microglial cells have a very specific and important—and helpful—reason for mounting an immune attack on the brain whenever the body falls ill. (And it isn't to fight the pathogen or virus per se—physical infections send immune signals to alert the brain, which in turn promotes neural inflammation, but the infection itself doesn't cross into your brain tissue.)

Microglial cells, it turns out, developed the ability to mount an immune offensive in the brain in order to help you heal, and to keep you and your family safe.

Let's go back to our tuberculosis-in-the-village analogy. Let's say you are one of the lucky few, and you slowly begin to recover from tuberculosis. You are still convalescing, but there are promising signs that you are turning the corner. You are going to live. Even as you continue to physically improve, however, you still feel horrendous. You feel a torpor-like fatigue, despair. You have impaired psychomotor function (it's hard to move, or lift your arm to brush your hair). You feel a nameless dread and malaise you can't shake. You want to curl up into a little ball and keep the covers over your head and rest, for what feels, at that moment, like the rest of your life. And that's because, even as your body begins to successfully battle the infection, microglia continue to spew forth inflammatory chemicals in your brain, causing changes to neurocircuitry that change your behavior too, in ways that will likely prompt you to experience a complete loss of interest in life—or anhedonia.

In other words, in addition to the physical symptoms that accompany tuberculosis, you are now also exhibiting what physicians call "sickness behavior." Even as your body continues to improve, your brain feels groggy. You still feel too tired to get up and wash your face, to dress yourself. You feel depressed, unmotivated, fatigued, and sleepy, and find it hard to concentrate.

So you mostly keep to your bed.

And the reason you feel so depressed and unmotivated is, once again, thanks to microglia. When you are sick, microglia crank up inflammation so that you don't feel like engaging with life.

And that, it turns out, is a pretty neat evolutionary trick.

Even as you start to recover, you still don't feel like moving, socializing, or engaging in activities that used to interest you. And that means that your immune system can leverage all of the body's resources for healing. With this more robust immune response, you get better faster. This also helps your kin: By staying in bed, you're less likely to spread your germs around. So your children and siblings and cousins are more likely to survive too, which allows them to pass the genes you all share along to the next generation. Your social withdrawal also minimizes the likelihood of your being exposed to additional infectious agents in the world outside.

By the time you feel clear-headed and well enough—mentally and physically—to go back to your normal day-to-day life of helping with the harvest, or selling market wares, you no longer need to conserve all of your energy for healing.

If you're a child and you survive, you get to grow up, have children, and pass your heartier immune response to this pathogen along to them, which will help them survive in the future.*

---

* Sickness behavior in response to pathogens has played a critically important role in our evolution as human beings; interestingly, many other animals, including social birds, do not exhibit sickness behavior. They can appear to be perfectly normal until the day they keel over.

Just a short period of microglia-enhanced depression, aka "sickness behavior," could save your life, and the lives of your family.

This is, as we all know, a classic tale of natural selection.

But what's entirely new is the recognition, among neuroimmunologists, not only that our immune systems evolved alongside microbes and pathogens, but that this coevolution conferred some benefit to our ancestors because of the fine-tuned glial immune response in the brain, which influenced our social behavior in profound ways.

Our ancestors' immune systems became "highly educated" to respond in smart social ways to a pathogen-rich environment. And it is microglia that gave us much of our evolutionary advantage in what was once a highly microbial and pathogenic world.

Meanwhile, in our twenty-first-century urban and suburban settings, we don't face as many threats from pathogens. Infections like tuberculosis are rare. (And, as of this writing, happily, we seem to be keeping a lid on plagues and pandemics.) We live in a much more sterile environment in general too. We don't meet up with the same microbes we once did: We don't sleep on dirt floors, or dig root vegetables from microbe-rich soil very often (even if you do grow your own vegetables, the quality of today's soil, for a host of environmental reasons, is more sterile than ever before).

Some of this is good news, and some of this is not such good news for us. Many of the evolutionary microbes that evolved alongside us in our natural environment kept our immune response busy and active in healthy ways too.*

---

* This is why one treatment for autoimmune disease involves what's known as *helminthic therapy*, in which a physician purposely introduces harmless intestinal worms called helminths into a patient's intestinal tract. Human beings coevolved with helminths in our natural environment, and our immune system recognizes them as old and familiar foes—and knows just what to do with them. In modern medicine, physicians use the presence of helminths to redirect a confused, overvigilant immune system to attack the worms instead of attacking a patient's own tissue, often helping to relieve symptoms in some autoimmune diseases and asthma. In animal studies, when newborn rats are exposed to bacterial infections, their levels of inflammatory cytokines

Too Clean and Too Dirty at the Same Time

We may be living in a world that's no longer rife with old familiar pathogens, but that doesn't mean that we're living in an environment that's "too clean."

In today's world, our immune system isn't encountering the accustomed nature-made pathogens, microbes, and invaders our ancestors came to know. But at the same time, we're inundated with new and completely unrecognizable man-made foreign invaders. And these are bombarding us from every direction. Which means we're living amid a very different backdrop of potential "threats" compared to those we evolved with. We're coexisting amid a chemical soup of manufactured environmental toxins: eighty thousand chemicals that have never been tested on the human immune system, much less on the brain's immune system. Nevertheless, all these chemicals are EPA-approved for use in items we come into contact with every day: flame retardants in furniture and carpets, dioxins in car exhaust, endocrine disrupters in cosmetics, plasticizers in baby toys, and toxic pesticides sprayed on our home gardens, farm crops, and agriculture, to name a few.

It's not just the air we breathe or what we slather on our skin that's problematic. Our diets have changed. Compared to our ancestors, we're ingesting a fair amount of processed foods full of additives, preservatives, and other artificial ingredients that may further confuse our immune system.

And many of these twenty-first-century triggers can cause our inflammatory response to go on high alert—poised for an all-out defense.

To use our villager analogy, it's as if your village was accustomed to ongoing skirmishes between your village and nearby warring fiefdoms. You and your comrades knew exactly how to respond to those factions.

Only now, suddenly, you're getting attacked by a new style of warfare—

---

rise, sparking microglial-led neuroinflammation. These same rats develop cognitive, learning, and memory problems later in life. But when researchers introduce helminths, newborn rats exposed to infection no longer develop cognitive disorders. The introduction of helminths protects the babies' brains by preventing microglia from entering an inflammatory state.

bombs and tanks you aren't prepared to defend against—from every direction.

Your immune system simply can't keep up with it all.

When Microglia Go Haywire

When your immune system is faced with endless triggers it can't recognize, your T-cells, white blood cells, and microglia can get overwhelmed. You might think of all these cells, which are struggling to respond to every millisecond of modern life, as being in a highly confused state: "What's happening here? Is this safe or not safe? Is this friend or foe? Should I respond to this threat? Or not?"

Suddenly—with no evolutionary playbook to go by—the immune system might just start responding all the time.

And that opens the door for the immune system, and microglia, to make more missteps, going haywire, promoting more inflammation and more disease.

Suddenly, the signals that Jony Kipnis talks about—which must be passed on in a perfect, exquisitely coordinated dance between the body's T-cells as they send alert messages through the brain's lymphatic immune pathways to the brain—start to break down. Errors abound.

In the body, when the immune system gets overwhelmed and signaling mistakes occur, our immune cells can begin to attack the body's own tissue and organs. In the brain, when the immune system gets overwhelmed, or microglia receive faulty signals, they start to spew out inflammatory chemicals and eat away at the synapses that give us mental stamina, hope, joy, and clarity of mind.

It's all one system, connected by a brain-immune superhighway, and when the body is overwhelmed, the brain can become overwhelmed too. The same environmental toxins, chemicals, and processed food diets that are catalysts for disease in the body can trigger microglia to launch an immune attack against the brain.

That's a lot to take in. And here, I think, is where things get even more intriguing. This information provides us with an entirely new lens through which we can make better sense of today's epidemic of adolescent and adult psychiatric disorders.

But not in the way you might think.

Depression and Anxiety: A Social Pathogen Epidemic?

According to Charles Raison, M.D., a professor in the human development and family studies department in the School of Human Ecology and department of psychiatry at the University of Wisconsin, Madison, and one of today's foremost neuroimmunologists, the way in which we evolved alongside microbes and pathogens matters a great deal in terms of understanding today's skyrocketing epidemic of depression, anxiety, and suicide.*

As Raison explains it, our immune system's coevolution alongside pathogens has led to what he calls an "evolutionary mismatch" with modern life—in several ways.

To extend our feudal village analogy, let's say you are walking along the road near your village, bringing home a rabbit you've just trapped for dinner. Along the wooded road you meet a wolf. In the first split second that you spot the wolf, your body's inflammatory stress hormones spike. You enter "fight-or-flight" mode—so that you can either run from the wolf or fight him off.†

---

* This new understanding of the evolutionary link between depression and immune response is largely due to the research of Raison as well as Andrew Miller, M.D., director of the behavioral immunology program in the department of psychiatry and behavioral sciences at Emory University.

† This is how the stress response works: As you fight the wolf, your sympathetic nervous system kicks in. Adrenaline courses through your body and brain: Your arms and legs become stronger, faster. Let's say you win the battle. Whew. By the time you are warm and safe again in your cottage, telling the story over a dinner of rabbit stew, your stress response begins to return to normal as your parasympathetic nervous system kicks in. That's how we evolved, over time, in the face of emotional stressors, in order to stay safe in a predator-filled world: going in and out of the two halves of the "stress cycle,"

Such fight-or-flight interactions spark a physically empowering inflammatory response. But, says Raison, your levels of inflammatory stress chemicals also rise because high-conflict interactions are very likely to lead to injuries and open wounds. "Across evolutionary time, stress has been a very reliable indication that one might be at increased risk of infection or death," he says.

After all, if you fought off a wolf, it was likely to claw you or bite you. If you fought another villager, he might very well bash you in the skull. And it was almost certain that infectious pathogens would enter those wounds. So, when facing conflict, your immune system went into overdrive to help ward off any pathogens that were a likely by-product of such stress-laden altercations.

Similarly, if you faced social stress in your village or tribe, that also signaled your body to produce a heightened inflammatory immune response. And that's because, across evolutionary time, social stress often led to getting into a fight, which could lead to wounds and infection, or to being ostracized, which in turn meant you'd lack food, shelter, and the protection of your family or tribe. You'd be exposed to the elements, predators, or hostile members of other tribes—unprotected—and therefore many times more likely to be wounded and at the mercy of infectious pathogens. Just the perception that one was under stress served as an early warning sign to prick up the immune system to prepare for infection.

Only now, says Raison, "When stress activates inflammation in the modern world, it's doing what it evolved to do, but, very often, it's doing it incorrectly, and to no good purpose." In modern life—versus life for our "villager"—we aren't usually coming into contact with wolves on our path or skirmishing with fists or weapons with other villagers. And that, he says, creates "an evolutionary mismatch. Most of the stressors we face today that activate inflammation no longer pose a threat of increased infection risk, so we are paying the price of having inflammation but without any compensatory benefits."

---

becoming hyped-up, stronger, and more vigilant when we needed to be, and then, as a stressor passed, coming back into a healthy state of relaxation and homeostasis.

Current studies tell us that in today's world, even a perceived psychological or emotional threat—like thinking about a big bill coming in the mail, or replaying an argument you had with your friend, your boss, or your spouse hours after it happened—can set off a physical inflammatory reaction. In modern life, says Raison, we are continually engaging our stress response and pricking up our immune system without relief.

Now, add to this that we have a second evolutionary mismatch: We are no longer exposed to a whole host of microorganisms and parasites that we emerged through time with. "Multiple facets of the modern world, from antibiotics to refrigeration and paved surfaces, have reconfigured our relationship with the microbial world in ways that have reduced our contact with a wide range of microorganisms," says Raison.*

In today's less microbial world, our confused immune system—no longer engaged in its familiar dance with recognizable pathogens and microbes—becomes hyperbusy looking around for something familiar to fill this void. And social-emotional threats and stressors, Raison argues, serve that purpose.

Which means that in the modern world, microglia are responding to emotional stressors as if they are biological pathogens.

Suddenly, we have a big problem. Our confused immune system sees these constant twenty-first-century stressors as a pathogenic threat and chronically releases, says Raison, "inflammatory cytokines that impinge on neurotransmitters and neurocircuits."

---

* Raison points out that many of these microorganisms were helpful and had powerful anti-inflammatory effects when they came into contact with our bodies. For instance, one such microorganism is *Mycobacterium vaccae*, an anti-inflammatory bacteria which, says Raison, "has anticancer effects and probably antidepressant effects as well." When researchers treated mice with *M. vaccae*, it reduced how primed microglia became to cause inflammation, while increasing anti-inflammatory factors in the brain's hippocampus. *M. vaccae* also prevented stress-induced increases in anxiety, while reducing anxious behaviors. This suggests that *M. vaccae* mitigates the neuroinflammatory and behavioral effects of stress on the brain, helping to impart stress resilience. However, in our modern world, we no longer come into contact with many of the good and the bad microbes that we evolved with across evolutionary time. And that is also changing our immune system's response to everyday stressors and toxic stressors in ways that we are just starting to understand.

· · ·

Okay, back to our imaginary village one last time. If you were deathly ill in your village hut, and the microglia in your brain responded to the infection in your body by causing changes to your brain circuitry that made you want to pull the covers over your head—well, that protected you, your family, your children, and your unborn children too.

In a similar way, if you feared you were being left out socially or excluded in some way, or if you were gearing up for physical conflict with someone, or to fight another tribe, or predators, your enhanced immune response would help, *temporarily*, to protect you.

That's the way the stress response is supposed to work: We enter a state of fight-or-flight and produce a heightened immune response to deal with an emergency, and after the threat has passed, our immune system returns to homeostasis. In acute circumstances, such as a physical fight, that inflammation is good; your body gets ready to fight off pathogens should you be wounded. After the fight, you relax. But in today's world, social stressors—perhaps especially social media stressors—can be constant, and levels of inflammatory cytokines stay elevated, potentially triggering microglia in ways that can contribute to depression.

In today's world—in our digital information age—as microglia chronically hyperrespond to emotional stressors as if they are pathogens, the stress response is always "on" and never has the chance to calm down and return to homeostasis as it should.

And suddenly, what was once a helpful evolutionary response—helping to keep you safe in response to the pathogens around you so that you and your loved ones could enjoy a better, safer, healthier life—no longer serves a helpful purpose.

Quite the opposite. When microglia misinterpret "threats" and begin to denude the synapses we need to get out of bed, to take care of ourselves, and to feel good enough about ourselves and our lives to want to engage in the world—this isn't helping us to survive the stressors we face. Now it's sabotaging us, keeping us from taking the steps and actions we need to take to help ourselves recover and thrive.

As Raison puts it, this new brain-immune response to social stressors

as if they are biological pathogens leads to "depressive and anxious behaviors that are poorly suited for functioning in modern society."

It makes sense that as we continue to outpace our immune system's evolutionary ability to keep up with twenty-first-century toxic stressors, the
rates of brain-related diseases are continuing to rise.

It's another simple equation of A + B = C.

A (Brain-related disorders are circuitry disorders generated by
microglia-neuron interactions.)
+
B (Microglia, in response to an evolutionary mismatch with
modern triggers in our environment, are interacting differently
with neurons, eating synapses, spewing out inflammation.)
= C (Microglia-led neuroinflammation plays a significant role in today's
escalating rates of brain circuitry disorders and psychiatric diseases.)

We might, for instance, apply this new calculation to better understand
today's escalating rates of depression among young and midlife men and
the social stressors they face. That is certainly an example of this theory.
The economic uncertainty many men face amid a dwindling middle class,
loss of skilled labor jobs, the lack of social safety nets such as college access,
retirement, and health coverage—all in the face of the cultural expecta-
tions we place on men that they should be able to be strong, never crack
under stress, and provide for a family—are enormous and often unrelenting
societal pressures. All of these social stressors, experts universally argue, play
a role in driving rising rates of depression, addiction, and suicide in men.

Girls, Disrupted

However, since Dori Schafer has brought up today's epidemic of depression, anxiety, and mood and eating disorders facing adolescent girls, let's
go back to girls as our primary example to help illustrate this hypothesis.

When Heather, for instance, looks back at her daughter's high school years, she describes them as a time of high-octane stress, even peril. In middle school and high school, Heather tells me, Jane, like most girls, was inundated with negative and/or sexist messages on social media. "Starting in middle school, the girls compared themselves to each other nonstop on social media. Jane would show us stuff girls were posting—hating on themselves, or putting up these stylized, unrealistic images of themselves." In retrospect, Heather says, "I wish we'd limited her social media time more. During those years, she developed an eating disorder and an anxiety disorder. I *know* it had a negative effect on her."

Marinating in the daily Insta, F'Insta, or Facebook Effect can be demoralizing for an adolescent who is trying to figure out how to create her own unique identity while also trying to fit into the girl tribe. According to Raison, this is a "classic evolutionary mismatch." In hunter-gatherer times, he emphasizes, "signals of impending ostracism needed to be taken very seriously because if you were rejected you really would die. Hence the horror that feeling ostracized or left out, even on social media, engenders today, even though these feelings are unrealistic in response to the actual situation" that one might face now. Trauma researchers who study the effects of adverse childhood experiences (ACEs) on the body and brain now cite social media—as well as academic stress—as a major adverse childhood experience for kids growing up today.

Heather, as a teacher, has seen how, from puberty on, girls are surrounded by media-delivered messages that they should measure up to some warped image of effortless physical female perfection. It's not as though our girls don't see the biological implausibility of it. If you don't fit the impossible norm—curvy, beautiful—if you are too fat, too flat, or too skinny, you don't measure up. If you do fit the impossible norm, men will drool and objectify you. Just being female means you may get sexually harassed or assaulted—stats tell us this is true, and headlines are full of women being sexually assaulted by powerful men. If you speak up, doing so can come back on you too. Whether you are physically desirable or not, you are always in fear of doing something wrong, and in some way being ostracized for it. This cultural sexism—this chronic perceived threat—gets inculcated into the psychic trauma that many girls live with every day.

Again, boys face their own version of being bombarded by messages pushing impossible standards for acceptable maleness. Some boys grow up with an outdated, even suffocating idea of masculinity that associates manhood with physical strength, aggression, and power over others, and come of age feeling that they can't express tenderness, fear, or grief without being emasculated. This can lead boys—especially if they are bullied at home or at school for not being tough enough—to feel ostracized and isolated—not part of the "tribe." Some boys may withdraw, or even feel enraged.

Meanwhile, both girls and boys, Heather points out, putting on her teacher hat, are also "constantly fretting online about what college they will get into—obsessing over whether their grades, activities, sports, and scores are good enough compared to other kids. And they fret about what's going on socially too. They can see on social media whenever they're being left out. Feeling you don't measure up or aren't 'good enough' to be included takes on a life of its own. When something happens, there is no fresh start the next day." All this, says Heather, "makes kids lose perspective; instead of seeing something difficult that's unfolding as being a temporary, negative event, it takes it all to an entirely new toxic level. If you can't get away from a sense of shame or guilt or worthlessness, you start to internalize those feelings and believe that *you are the problem*."

What a way to come of age.

At the same time, the developing teenage brain is wired to care deeply about social connectedness, which means that the threat of losing social connection, or being shamed by a vocal, texting tribe of other girls, or by society writ large, carries enormous emotional impact, delivering a double wallop to the immune system.

All of this sets girls up to swim in a body-shaming, self-hating, emotion-suppressing, fear-primed, self-blaming toxic emotional soup from a very young age. And it places young girls in a constant biological fight-or-flight (or freeze) state. If you're getting the constant message that with one false step you'll be cast out of the girl tribe—and on the other hand you're getting the message that girls and women in general aren't safe in a sexist world—then you don't feel safe on any level. You're not entirely safe among your peers, and as a girl, you're not safe in our world.

And that is a chronic social stressor.

The time girls spend on social media and scrolling through news headlines on their computer or smartphone becomes a lot like sitting in front of a bullhorn that's screaming out to the brain *"Watch Out! Incoming Danger! Pathogen Alert!"*

Not surprisingly, Johns Hopkins researchers reported in 2016 that the more time a teenage girl spends on social media platforms, the more likely she is to develop depression, anxiety, or a mood disorder. In fact, the use of Facebook and other social media sites almost seems to predict mental health disorders in young adults. Teens who spend more than five hours a day online are 71 percent more likely to suffer from depression or consider suicide than are teens who spend less than an hour a day online. There are stronger associations between social media use and depressive symptoms for girls than for boys. Researchers believe this may have to do with the fact that girls are more likely to use sites like Snapchat and Instagram, which are based on sharing photos of and receiving comments on one's physical appearance—body, clothes, hair—coupled with the fact that girls, in general, are at greater risk for depression than boys.

In 2012, 50 percent of Americans had a smartphone. By 2015, 73 percent of teenagers had smartphones. During this same three-year period, the teen suicide rate skyrocketed.

These studies on teen use of social media and depression show correlations, of course, not causality. It might be that the direction of causality goes the other way: Teens who spend so much time on social media are already depressed and self-isolating, and they are seeking solace or comfort in the wrong place: online. We don't know. However, one 2019 study from the American Medical Association argues that electronic communication and digital media—as well as less face-to-face interaction—may indeed be causal in today's rapidly rising rates of teen mental health disorders. Researchers found that depression, suicidal ideation, and suicide attempts had increased significantly among teens and young adults since 2011, but this trend was "weak or nonexistent" among those who were twenty-six and older—the generation that came through their teen years just prior to the age of social media.

Either way, put all this together, and it may be that the increased pressures teenagers like Jane face both from social media and academic stress trigger the brain's microglia to go overboard, express inflammatory factors, and eat away at synapses. Social stressors are not the same as microbes and pathogens. But to the brain, maybe they are.

Adolescent girls' terrifyingly high rates of depression, anxiety, and eating disorders warn us that the psychosocial factors girls face may be particularly malevolent. Social media may be functioning as a nonstop delivery method for social pathogens—at the brain's most vulnerable time. It's a brain-body bidirectional feedback loop—with social stressors pricking up the overactivity of a girl's immune system.

Which means that our girls may be becoming more anxious and depressed from a modern kind of social plague.

This sounds dramatic, but the science tells us that it's true.

No wonder it seems as if our girls are falling sick in droves with a strange, highly contagious flu. In a way, they really are.

Again, the brains of adolescent girls are not the only brains in which microglia are going haywire. Poor diets and environmental chemicals are confusing the immune system, while modern social stressors (economic uncertainty, loss of community, increasingly toxic political "tribalism," the Facebook Effect, and a lack of societal safety nets) are simultaneously acting as social pathogens, in many different groups in different ways, potentially amplifying microglial reactivity and contributing to rising rates of depression, cognitive decline, and other brain-related disorders from puberty to old age.

But there is good news on the horizon too. Thanks to a number of remarkable, fearless neurobiologists, who are treating patients' brain disorders based on the microglial universal theory of disease, we are now increasingly able to intervene with new empathy, understanding, and treatment options and offer patients new possibilities of lives reclaimed.

EIGHT

Brain Hacking

THE WALLS OF DR. HASAN ASIF'S MEDICAL OFFICE ARE DECO-
rated with photos of sand mandalas interspersed with his diplomas
and certifications from the American boards of psychiatry and neurology. On one wall there hangs a striking Asian ink-wash painting of what
appear to be floating trees. At the base of each trunk, intricate spiderweb-
like roots shoot out and burst into flower with dozens of tiny pink cherry
blossom buds dangling at their tips. But when you look closer at these
delicately painted trees, you can see they are actually artistic renderings
of neurons as they might appear under the microscope, splaying their
long tentacles across the microscope slide.

It is as if the artist has captured the job of neurons in a single Zen-like
image: to help our brain flower so that we are deeply rooted, and strong,
in this life.

And indeed, once we start to wrap our minds around the new scientific
understanding that brain-related disorders are set in motion, in large part,
by dysfunctional microglia-neuron interactions, we can also focus our
attention on how to promote the recovery of wayward microglial assassin
cells so that they transform back into helpful cells that restore beautiful
synapses and brain networks working for us rather than against us.

And this, it turns out, is exactly what Hasan Asif, M.D., does.

A neurotherapist and founder of the Brain Wellness Center—which has offices in New York City and in Bronxville, New York—Asif works in a new area of neurotherapy often referred to as "brain hacking." Brain hacking techniques utilize neuroengineering tools to help coach and stimulate underactive or overactive brain circuitry and brain waves to function in healthier ways. Asif is one of a number of early adopter clinicians bridging the gap between cutting-edge lab science and patient care, taking the best of brain hacking methods, and research on rebooting microglia and neurons, straight to the front lines of patient suffering.

In his clinic, Asif, who also serves as a psychiatrist at New York–Presbyterian Lawrence Hospital, has been primarily utilizing a brain hacking method called transcranial magnetic stimulation (TMS) for years, in an effort to help patients with hard-to-treat major depressive disorder and panic disorders—patients like Katie, whom we first met in chapter 2.

At fifty-two, Asif is slim but not skinny, with thick, graying black hair neatly combed back from his square, symmetrical face. His eyes are dark and enormously responsive when you meet him for the first time, as if he suspects that everyone who enters his office is bearing up under an invisible cargo of pain, one he's ready to help you put down, once you get beyond the cheery hellos and how-are-yous.

Asif grew up in Pakistan and trained as a psychoanalyst before he became a psychiatrist. He did his postgraduate work in the United States, at New York Medical College in Valhalla. It was 1990, and psychiatry was consumed by the idea that psychiatric disorders were chemical disorders caused by deficiencies of serotonin, dopamine, and other neurochemicals. Psychiatric hospitals and practices were tiny boomtowns (and still are) for pharmaceuticals. "As medical trainees, we were surrounded by drug reps," Asif says.

But Asif began to take note that the antidepressants they were prescribing were working well for less than half of patients.* And when they

---

* Studies show that, statistically, one-third of patients with psychiatric disorders fail to respond to any antidepressant treatment.

did work, they didn't work right away. It often took weeks for medications to begin to help patients feel a modicum of relief. Often the positive effects wore off with time, requiring physicians to increase doses and add more medications—which in turn led to unpleasant side effects: sedation, weight gain, brain fog, sleep disturbances.

Dr. Asif is clear that he is in no way against pharmaceuticals; they are an important part of treatment. Still, he recognizes that medication leaves many patients, like Katie, living what can feel like little more than half a life.

As a medical fellow, Asif began asking questions about psychiatry's increasingly laser-like focus on medication. Once, he recalls, during his medical training, a young girl who had just been forcibly separated from her parents came into the hospital suffering from anxiety. A colleague immediately began dosing her with SSRIs. But Asif was curious. He wondered, "How is *that* serotonin? Did that mean that if you were having anxiety very early in life, as a kid, you didn't have enough serotonin?" What, he wondered, about the fact that she'd been emotionally traumatized?

"I began to feel that I would have to make my own bridge between psychoanalysis and biological psychiatry," Asif says. "I was interested in both a patient's whole, lived experience and what was going on, at a structural level, in their brain." But, he says, "The level of discussion taking place at that time about what caused these imbalances in any given individual's brain was very disappointing. I felt we were failing so many patients."

As a young resident, Asif began keeping a small notebook in which he methodically attempted to chemically profile people. Was it possible, he wondered, that developmentally, certain areas of the brain that were more vulnerable in early life were altered by loss, which later translated into depression and anxiety, and this imbalance was then helped by serotonin?

"I thought, okay, perhaps those who did not have secure attachment with their caregivers, and who suffer from attachment issues, benefit most from serotonin," he explains. "Similarly, perhaps a different area of the brain is affected in attention issues and that is why these patients do better with dopamine."

He admits, in retrospect, that his theory was flawed, but as a new analyst and psychiatrist treating patients, he was struggling to take the vastly oversimplified notion consuming his profession—that there is a drug for every chemical imbalance that will fix the patient's symptoms—and make better sense of it in terms of a patient's early life experiences and real-time stressors. He felt there was some missing link that psychiatry as a whole wasn't seeing.

"My profession had become the child of a bad marriage between the biological and the psychodynamic, thrown into the custodial care of pharmaceuticals," he says, tapping his desk with his finger for emphasis. "And the parents didn't respect or speak to each other."

Asif's own experience, growing up, also made him intuitively doubtful of serotonin as a sole solution for intractable sadness or anxiety. "My father died suddenly when I was eleven," he says. "It changed who I am. So I suppose I was also trying to understand my own path of grief, and how it could have so radically influenced every aspect of my life thereafter."

Asif's sense of being unanchored by loss fueled his desire to better understand his patients and help them in novel ways, often at the risk, when he was a young physician, of being dismissed by his peers.

"So, how did you start to close that gap between psychoanalysis, human suffering, and neuroscience?" I ask.

Peering Deep Inside the Brain

In 2007, Asif began using a qEEG brain scan, or quantitative electroencephalogram, to try to find out more about what was going on inside the brains of his patients.

In a qEEG, a patient wears a cap fitted with nineteen small holes. Small, flat disc-like sensors, or electrodes, are attached through these holes to a patient's scalp and connect to wires that in turn deliver data into a computer system. The electrodes measure tiny pulses of electrical activity in different key areas of the brain by reading alpha, beta, theta, and delta brain waves. Alpha waves are associated with relaxation and calm; beta waves with mental focus (such as when taking an exam); theta waves with a dreamlike, meditative state; and delta waves with falling asleep. Measuring these different types of electrical brain waves indicates which areas of the brain are functioning normally and which are not. This data measuring the electrical activity of the patient's brain is fed into a computer, quantified by an algorithm, statistically analyzed, and interpreted into a patient's "brain map." This brain map is then compared to an extensive national database of normal, healthy brains.

The first time Asif used a qEEG as a diagnostic tool in his practice, he felt he was staring at one of the missing links he'd been searching for. He spent the next two years "just trying to get good at qEEG before I treated patients using it," he recalls. Once he felt he had mastered reading brain maps, he offered qEEG neurofeedback—a brain-training method that helps patients learn how to increase or decrease their brain wave activity in the parts of the brain related to their symptoms.

In qEEG neurofeedback, after having had electrodes glued to their scalp, patients play a computer game while trying to increase or decrease brain circuitry activity in areas of the brain that are over- or underactive. For instance, a patient may see cougars running across a screen. When the patient's brain is showing slow brain wave activity in an area correlated with depressive thoughts, the cougars hop very slowly. The more the patient focuses her mind in a way that causes that underlying circuitry to fire up, the faster the cougars race across the screen. Another patient may hear a series of enjoyable musical tones, which turn into unpleasant, chaotic noises if he stops focusing on training his brain wave activity. The practitioner uses these sounds or video images to give the patient positive or negative feedback in real time, depending on whether the desired brain wave activity is being achieved. Over time, poorly regulated brain wave patterns become better regulated. As brain circuitry begins to be retrained, behaviors begin to change.

Asif felt that using qEEG neurofeedback was unquestionably helping his patients.* But the process was, in difficult cases, slow. For many patients with psychiatric disorders, self-neglect is a major component of their disease. And for patients for whom daily life was already very hard—depressed patients who were also slogging through the physical symptoms that resembled "sickness behavior" and who lacked the motivation to self-care—neurofeedback could seem like overwhelmingly hard work at a time when taking steps to care for oneself felt like a mountain too high to climb.

Unlike many neurofeedback practitioners, who let the computer games do the treating, Asif used his considerable training as a psychoanalyst too; he talked with his patients during neurofeedback. He asked probing questions about their childhood, parents, love relationships and marriages, their children, extended family dynamics, their work life, and any other key life stressors. What made them feel terribly sad? When did they remember feeling happy?

Like any good therapist, Asif didn't take long to see what issues were emotionally sticky for any given patient, triggering their anxiety, rumination, fear.

And while he talked with patients, he kept checking in on what was happening on their qEEG brain maps. "If I asked a patient to think about a particular past trauma that happened when they were growing up, certain areas would be more activated," he explains. Asif quickly saw how entrenched emotional patterns of loss, regret, trauma, or fear, had, over time, dramatically affected specific areas of brain circuitry.

For instance, he says, "I could see that neurocircuitry changes in depression were different from those in patients with anhedonia, or a loss of interest in life. Or, if I saw the brain going into a freeze state or a slow

---

* In 2007, studies on the efficacy of neurofeedback were quite mixed. Today, evidence increasingly indicates that neurofeedback can be an important adjunct therapy, helping to improve mood dysregulation and stability in some patients with PTSD, improving symptoms of ADHD, and relieving symptoms of generalized anxiety disorder in some patients. In one 2018 single-blind randomized controlled trial of patients with major depressive disorder, 38 percent of patients who underwent neurofeedback went into remission, and their depressive symptoms decreased by 43 percent as compared to symptoms prior to treatment.

wave state, in the right parietal and right insular lobes, in response to questions I asked about growing up, I could predict, before the patient told me, that this was an individual who had experienced a lot of developmental trauma, very early in their childhood," he says.

Over the next decade, Asif brought in newer neurotherapy tools as neuroengineers developed them. He was particularly excited by the discovery of a novel way to peer deeper into the brain by using a computer program called sLORETA, or standardized low-resolution electromagnetic tomography. The sLORETA program synthesizes the information relayed by the electrodes on the patient's scalp and takes it one step further than a qEEG brain map. Whereas the qEEG consolidates information from EEGs, which capture quantitative and qualitative data on how brain waves are functioning, low-resolution electromagnetic tomography uses this qEEG data to produce a more dynamic, live, three-dimensional, visual image of the patient's brain in real time. This live, dynamic brain scan—which looks a lot like a live PET scan—utilizes a color scale that runs in shades from cool to hot. If there is too much neural activity, as compared to a normal brain, that brain area will glow red. If there is too little activity, the area will glow blue. Areas where the brain is humming along appear green. It's like watching a live movie of what's happening inside someone's brain.*

Asif used these dynamic live brain scans, as soon as they became available, in his work with thousands of different patients with varied psychiatric, behavioral, and cognitive disorders. And by noting where circuitry appeared to be abnormal during a patient's different emotional responses, he felt he was isolating "a new biomarker for disease that psychiatrists were largely missing."

After assessing thousands of patients' live, dynamic 3-D brain scans, Asif knew, for instance, that more brain wave activity in the temporal lobe was related to more rumination—negative thoughts that a patient

---

* Some TMS practitioners use magnetoencephalography, or MEG, to record brain activity and create a 3-D live image of the brain, which shows where and when key areas of the brain are over- or underactivated. The main difference is that an EEG relies more on reading electrical activity, and an MEG relies on magnetic fields (somewhat like a PET or MRI brain scan) to read brain activity.

just couldn't let go of. When patients were dissociating, or denying and repressing emotions and ignoring the physical sensations that these emotions brought up in their bodies, he saw major deviations in the insular and parietal lobes of the brain. He knew there was some important emotional awareness that this patient was not allowing himself or herself to feel. Similarly, feelings of panic announced themselves as changes in synaptic activity in the limbic and dorsal prefrontal cortex areas of the brain.

He began to think of these patterns as "neuropsychoanalytical circuits" that could allow you to literally see the unique, fine-grained suffering of each patient's pain.

It was in 2014 that Hasan Asif first became interested in using transcranial magnetic stimulation in his practice. TMS was an approach that had been in use in hospital settings for some time, including at the Mayo Clinic and Harvard Medical School. That year, the FDA approved the use of a specific TMS device, which had been developed by neuroengineers, for patients with medication-resistant depression. *

Multiple randomized clinical trials have backed up the efficacy of TMS as a treatment for major depressive disorder. In 2016, one of the pioneers of TMS research, Alvaro Pascual-Leone, M.D., Ph.D., from Beth Israel Deaconess Medical Center and Harvard Medical School, demonstrated that TMS can create rapid and long-lasting benefits in neural circuitry, restoring healthy brain circuitry in patients who — like Katie — have not responded to traditional therapies for major depressive disorder. † Some patients who experienced significant relief of symptoms after thirty treatments with TMS never experienced a relapse of depres-

---

* Following FDA approval of the Neuronetics NeuroStar TMS Stimulation device in 2014, the FDA has approved five other stimulation devices for patients with medication-resistant depression.

† Alvaro Pascual-Leone conducted his first double-blind study on the use of TMS in treatment-resistant depression, and published the results in the journal *Lancet* in 1996, more than twenty years ago.

sion, even many years later. Other patients who respond to treatment, says Pascual-Leone, "need a repeated course of TMS every six months or so. With follow-up treatments, they continue to derive benefit and can thus cope with their depression and carry on a full and fulfilling life."

TMS allows a practitioner to deliver very brief magnetic pulses to precise brain areas, noninvasively, from the scalp's surface. In the brain, these magnetic pulses induce an electrical current that can help modulate underactive or overactive neural circuits. These energy pulses are akin to what you might feel if someone were to tap you gently and repeatedly on the head with their fingertips. (To be clear, this is nothing remotely like electroconvulsive therapy, or ECT.)* These pulses can be

---

* ECT, or electroconvulsive therapy, has long been seen as a controversial treatment. However, recent improvements in treatment protocols and methods to ensure safety and efficacy, coupled with more rigorous clinical trials, have demonstrated that ECT treatment is effective in 70 to 80 percent of patients with drug-resistant depression. Although ECT has been used for decades, the underlying mechanisms by which it works in some patients have remained elusive. In 2018, German researchers found that the therapeutic effects of ECT appear to be the result of the bidirectional interplay between physical inflammation and microglial activation in the brain. Researchers measured levels of inflammatory biomarkers, or cytokines, in the cerebrospinal fluid and blood of twelve patients before and after they underwent a course of ECT treatment. Patients with higher inflammatory biomarkers for macrophage/microglial activation prior to ECT treatment were more likely to respond well to ECT. This indicates not only that microglia are involved in the pathophysiology of treatment-resistant depression, but also that the efficacy of ECT might lie in the way in which it causes a reduction in the activity of microglia and macrophages, especially in patients with higher levels of inflammation. In another 2018 study, researchers found that patients with treatment-resistant depression, especially women, who had heightened levels of C-reactive protein and other inflammatory biomarkers, were more likely to respond to treatment with ECT, as compared to patients without high levels of inflammatory biomarkers. In 2016, researchers found that in animal studies, mice that demonstrated microglial activation in the brain's hippocampus alongside depression-like symptoms after being exposed to chronic unpredictable stressors responded well to ECT treatment. And that was because ECT significantly altered the presence and activity level of microglia in the brain's hippocampus, leading in turn to an increase in neurogenesis, or the birth of healthy new neurons. Although ECT is thought to be successful in more than two-thirds of patients with treatment-resistant depression, it can cause short-term loss in verbal memory, language capacity, and executive function during the hours immediately following treatment. For this reason—and because of the invasiveness of the procedure—it is most often used as a last line of treatment.

delivered with pinpoint precision through the skin, scalp, and skull into areas of the brain, while a practitioner observes which areas are either overactive and lighting up too much (bright red) or are underactive (bright blue) on dynamic brain scans. When successful, these pulses can begin to reregulate the electrical firing pattern in the brain. Over the course of TMS treatment, which usually requires twenty-five to thirty very short sessions, this begins to change circuitry patterns. Areas of the brain that have not been functioning as they should start to work properly again.

Consider a patient like Katie, who has had thousands of hours of therapy over the past twenty-five years to manage her major depressive and panic disorders. Katie has been going to several appointments a week—EMDR therapy, talk therapy, trauma-based CBT, biofeedback, psychiatrists, nutritionists, as well as visits with integrative physicians—every week for several decades.

Thirty treatments, while a lot, begins to sound like very little in the scheme of decades lost.

Was it possible, Asif wondered, to use TMS to help bring his patients relief from their suffering—so that they might never relapse again? He underwent extensive training in TMS, and after investing tens of thousands of dollars in TMS neurotherapy tools, he began utilizing transcranial magnetic stimulation, in conjunction with qEEG brain maps, live dynamic brain scans, neurofeedback, psychotherapy, and medication changes when indicated, to help his patients.

With these new tools, Asif says, "We could begin to treat panic attacks, *live*, as patients were talking. And we could see how well a treatment was succeeding in helping them too.

"I felt so much empathy for the anguish my patients were in when they came into my office," he continues. "It was so clear, looking at their dynamic brain scans, that so many of them had been through a wringer of misery. It was as if the brain was screaming out loud, trying to shout at us: *Read me! Look, here's what's going wrong here, please help me.*"

By 2016, brain mapping tools had evolved even more.* Scientists were using these new scanning tools in the lab to better understand psychiatric disorders in the brain. The underlying circuitry in different subtypes of depression became apparent. In animal studies, researchers could discern signatures in brain scans that told them whether depression had been specifically stress-induced, leading to a kind of helpless behavior, or an inability to take care of oneself, when encountering stressful new situations. This helpless behavior was correlated with less activity across almost every area of the brain—in areas associated with motivation, processing emotion, learning, memory, and organizing thoughts and actions—but it also correlated with more activity in one tiny spot in the brain, known as the *locus coeruleus*, an area in the brain stem that moderates our physiological responses to stress and panic.

Patients with anhedonia, who were unable to experience pleasure even when really good things came their way, showed specific deficits in an area of the brain known as the posterior ventromedial prefrontal cortex. Suddenly, researchers had a template for less resilient brains versus more resilient brains.

It looked as though, in depression, almost the whole brain was shouting, "There is nothing you can do to help yourself, so don't try!" while another tiny part of the brain was shouting, just as loudly, "Something terrible is headed your way!" In other words, patients who were deeply depressed weren't, as friends, family, or co-workers might secretly believe, simply unable to pick themselves up, dust themselves off, and pull themselves out of it. Their brains were telling them *Just don't bother.*

Brain scans also became increasingly useful—though certainly not foolproof—in predicting the course of psychiatric disease in any given patient. Patients initially diagnosed with depression who demonstrated too

---

* In 2016, researchers at the University of California, Berkeley, created a new "brain atlas," which specified which areas of the brain responded to different prompts, words, emotions, and feelings.

little activity in the left midfrontal lobe still had a primary diagnosis of depression four years later. However, patients who demonstrated too much activity in the left midfrontal cortex at the time of their depression diagnosis often had a different trajectory of psychiatric disease. Four years later, many of these patients' psychiatric disorders had converted into bipolar 1 disorder.

The increased ability to understand dynamic brain scans helped practitioners to discern what was plaguing the mind in much the same way physicians use blood tests to diagnose physical disease.

At the Mayo Clinic, TMS treatments demonstrated promise too in helping patients with autism spectrum disorder experience more emotions, by correcting deficits in neural wiring in areas of the brain that had to do with executive functioning and spatial working memory.*

One important caveat: This is not to suggest that the goal of TMS, or neurofeedback, is to create brains that all fire alike, in some ideal pattern. These treatments aren't intended to eradicate the full spectrum of human experience, including the melancholy, sadness, creativity, and joy that give individual meaning to our inner lives and make us undeniably human. Every brain is unique in its own way. Rather, these modalities work to remove what's hindering an individual's ability to function in and contribute to the world, and to love, in meaningful ways without removing their uniqueness.

---

* TMS has been shown to be effective in helping patients who suffer from obesity. In one study, half the patients received fifteen sessions of TMS and half received a placebo treatment. Before and after the study, subjects gave stool samples to researchers, who analyzed their microbiota. Researchers also measured patients' biomarkers for different neurotransmitters and neurochemicals. After five weeks of treatment, subjects who received TMS lost more than 4 percent of their body fat, significantly more than the control group. And the TMS-treated subjects showed increased quantities of a number of beneficial and anti-inflammatory bacteria in their gut microbiomes—strains that are commonly found in healthy people who are not overweight and that are thought to stimulate and control physical cravings for carbohydrates and fatty foods. The control group showed no such changes. This is likely because, as we shall see in later chapters, strains of bacteria in the gut microbiome also influence the activity of microglia in the brain, and vice versa. These studies are not yet published, however, and will need to be repeated and replicated.

The Microglia-Neuron Connection: Making the Brain New Again

Meanwhile, the burgeoning field of glial biology was completely changing neuroscience. Which led to new questions: Exactly how does transcranial magnetic stimulation reboot microglial cells so that they stop gobbling up synapses and start helping the brain instead?

We know that when microglia go on the attack, they prune synapses. And the newest research on microglia shows that when they spark runaway inflammation, they destroy a special type of neuron in the brain's hippocampus that, under healthy circumstances, is remarkably capable of regenerating. Microglia begin to assassinate these nascent neurons as soon as they are born—which is, researchers think, the reason why the hippocampus shrinks so dramatically in depression and in individuals who've experienced developmental trauma.

Yet, when researchers can get microglia to calm down, these naturally regenerating neurons begin to repopulate the brain.

So exactly how do the pulses delivered in transcranial magnetic stimulation call off microglia so that new, healthy baby neurons can be born, restoring brain synapses—and, in a sense, making the brain new again?

It's a tall order—a bit like trying to call off a hit man right before a job and telling him you no longer want him to go through with the deed.

The brain is an electrical organ. We know that neurons, in order to do their job in the brain, must pass messages along to other neurons. A typical neuron fires an electrical pulse five to twenty times every second, connecting to as many as ten thousand other neurons. As neurons fire away in the brain, they release electrical pulses along the brain's synapses billions of times per second.

And microglial cells provide neurons with just the right nutrients and growth factors so that they can grow, stay strong, fire up, and broadcast the right messages across the brain. If microglia don't tend to neurons in this fashion, neurons cannot do their job, or fire, correctly. When neurons start to flail, they begin to overfire or underfire.

In brain scans, this overfiring or underfiring is captured by quantifying a patient's brain waves, such as alpha, beta, theta, and delta waves, which are moving either too quickly or too slowly.

Research shows that in addition to helping neurons grow and populate the brain, microglia also help maintain the electrical synaptic activity that manifests as brain waves and help regulate healthy brain wave activity. We are all constantly cycling through different types of brain waves, which are also associated with different mind states. When we are cycling through brain waves normally, we feel pretty great. When they are out of sync, we feel unwell.

Areas of the brain that appear to be offline in a qEEG readout of a patient's depressed brain, or in a dynamic live brain scan, can be seen as what researchers think of as "microglial-neuronal interactions" that are not firing properly. Any deviation from the norm in delta, gamma, theta, and beta waves alerts practitioners like Asif that these microglial-neuronal interactions are not functioning as they should.

Seen in this light, delivering a gentle electrical pulse that the brain recognizes as familiar makes a great deal of sense. Delivered correctly, this noninvasive TMS charge may have the power to reboot microglia, handing microglia and neurons a fresh, proper set of signals, so that now microglia back off, they stop devouring synapses and instead begin to enhance the growth of new neurons and support the activity of synapses again, as they were originally intended to do.

The whole microglia-neuron interaction is restored to its proper, healthy function.

Beth Stevens at Harvard helps me get to the bottom of what could be happening here.

"Microglia are constantly responding to neuronal and synaptic activity," she offers. Given that we know that microglia consume and remove synapses that are less active, it makes sense that "bringing circuits back online via techniques that stimulate neural activity can have therapeutic value. If neural activity interventions are working, then at the very least we can start to think about ways to help patients other than just throwing a bunch of drugs randomly into someone's brain," she says. "And that is progress."

"This dance between neurons and microglia is like a beautiful choreography," says Hasan Asif. "When one of the dancers is out of rhythm, brain circuitry becomes disregulated in specific networks. The dancers fall out of step. They can't communicate with one another coherently." However, he underscores, "When we get the brain back in rhythm by increasing or decreasing brain wave activity, and restoring underlying circuitry, we can see the patient's symptoms improve rather quickly."

And this is very hopeful news for the fields of psychiatry, psychoanalysis, immunology, and neuroscience, says Asif. "Because it shows us that these seemingly disparate fields are *one* field. And if we combine our knowledge, we will almost certainly be able to help more patients heal."

NINE

A Beleaguered Mind

'M SITTING WITH KATIE HARRISON ON A DARK, WORN LEATHER couch in the patient waiting room of the offices of Dr. Hasan Asif's Brain Wellness Center in New York City.

Coming here was, Katie tells me, a big decision for her in terms of time, travel, and expenses—but in terms of her health, it was a remarkably simple one. After we last met, she explains, she did a little thinking, and a little research. She felt it made sense for her to try TMS. She has tried everything else that psychiatry and medicine have to offer, and although TMS was once considered futuristic medicine, the research on its success in treating major depressive disorder and anxiety disorders is now well replicated and impressive. Plus, she tells me, her psychiatrist at home in Arlington, Virginia, agrees that she may be helped by TMS and supports her being treated by Dr. Asif.

So Katie has come up from northern Virginia to New York by Amtrak for a brain scan evaluation to see if she's an appropriate candidate for this therapy.

She's exhausted, she says, after traveling. The noise and chaos of being in train stations, interacting with strangers, navigating the fast-paced, clogged sidewalks of New York, and staying in a hotel have drained her. Katie's hotel is not far from Asif's office, but she was so tired and overwhelmed when she arrived that she took an Uber rather than walk two city blocks. "I almost turned around a dozen times," she says. "I would have, if I wasn't so hopeful.

"I know my family history, my own history," she continues. "I've seen my lab tests. My inflammatory immune markers are in the range found in patients who also often have major depressive disorder. If I had cancer, an oncologist wouldn't just treat my symptoms, he'd treat the underlying cause of the disease. I want to know what's going on with my brain's health. If my higher levels of physical inflammation are triggering microglial cells in my brain to zap out synapses—that has to be contributing to why I feel the way I do, and why I can't seem to get any better.

"In some ways I feel as if I do have a type of cancer; my life feels as if it is already over. If I had had this kind of failure rate with treating a physical disease, I wouldn't just sit back and do everything exactly the way I was doing it before. I'd see what the newest science has to offer. And from what I've read, TMS is helping a lot of people. So I want to see if Dr. Asif might be able to help me.

"I'm also here for Mindy and Andrew," she adds. "I know the toll it took on me to grow up sensing that my own mom was not okay. I want them to have a different future—and break this family cycle."

As Katie talks, I observe the same thing I noted the last time we met. As she shares deeply painful things, the expressions you'd expect to see reflected in her face, the inflections you'd expect to hear in her voice, are missing. It's as if her thoughts are not really connecting with her feelings. I can't help but wonder, will treating Katie's underlying brain circuitry change what psychiatrists would call her "lack of affect"?

Despite her resolve, Katie is nervous. "I can't stop worrying about what Dr. Asif will say," she says, her face gray, stiffening as if she's just remembered that she left a teakettle on the stove at home. "I am afraid he'll say he *can't* treat me, or that there is something so wrong with me that he *won't* treat me." Katie gives me a too-quick artificial smile as if to mask her concern.

I think of one thing I have not shared with Katie. In my discussions with Alvaro Pascual-Leone about his research on TMS, I learned that a patient's initial response to TMS predicts whether she is a good candidate, in general, for TMS treatment, and whether she is likely to respond positively to an entire course of thirty treatments. If the brain does not respond well early on, a patient is far more likely to relapse later.

After spending so many hours interviewing and talking with Katie, I feel like her advocate; I want her to find the relief she longs for, and that all the patients I hear from, who feel there is so little hope for them, deserve.

Ten minutes later, I'm seated in a corner of Dr. Asif's main treatment room, taking notes as unobtrusively as I can as he begins Katie's evaluation.

Katie is reclining in an oversized comfy beige leather chair. She's wearing a bright yellow cap that has nineteen small holes in it, through which the electrodes can be glued onto her scalp. Asif's team uses sticky washable paste to secure the electrodes, which connect via thin black cords to a row of computers. These computers will scan and interpret Katie's brain activity.

By the time they are finished hooking her up, wires extend from all over Katie's head like delicate spider legs. Beneath the bright yellow cap, Katie's pale, planar face looks scared.

Dr. Asif has secured a respirator belt around Katie's abdomen, and sensors to the first three fingers of her left hand, to measure her autonomic nervous system's activity: breathing rate, skin conductivity (sweat and perspiration), and heart rate.*

"Relax the muscles around your face, your forehead, especially around the temples," Dr. Asif coaches Katie as he begins her brain scan. Otherwise, he explains, the brain waves he's reading will be full of artifact from muscle tension. He talks to her gently, in a meditative tone. "That's better, much better," he cajoles, his slight Pakistani accent soft, almost lyrical.

The room is utterly quiet as he creates a map of Katie's brain. "Now close your eyes and think about something that makes you really happy."

---

* Asif credits his understanding of how to utilize autonomic nervous system equipment in helping to monitor and treat patients to his partner and fellow neurotherapist Aza Mantashashvili, M.D.

From where I sit, I can see the colors that represent Katie's brain on her live, dynamic brain scan begin to shift.

"Now think about something that makes you feel very sad," he asks a few minutes later. Again, within seconds, the color palette of Katie's brain imaging begins to change.

A few minutes later, he invites Katie to open her eyes again.

Next, Asif shows her the image of a sunset and records her brain's response. Then he gives her a warning that there will be an unpleasant image coming up. Suddenly, we see the World Trade Center engulfed in smoke and flame.

After that image disappears, Asif tells Katie that he is observing how well her body and brain wind down after stress.

This whole process—hooking Katie up to the extensive neurotherapy equipment, reading her brain activity, and producing a dynamic 3-D brain scan—has taken about half an hour.

And now—as Katie sits with electrodes protruding from the cap on her head, all of which feed data into the computer—we see, twirling slowly in front of us, on the large computer monitor, a three-dimensional scan of Katie's brain—every lobe, from every direction.

Asif begins to decode the patterns he discerns in Katie's brain circuitry in a layman's language that Katie and I can easily grasp. As he talks he points to the various areas.

He puts his finger on the image of Katie's left temporal lobe, which is lighting up a bright red. "We can see on the left side that your brain is highly active, far more active than on the right, where we can see that the brain is not as active as we would like it to be." These are, he underscores, significant deviations from the norm. Asif slides his fingers across the computer's monitor, to the right side of Katie's brain. "Here we see a significant deviation in the right insular region, the right parietal lobe, and the right prefrontal cortex: a slow wave predominance in areas having to do with your affect regulation."

This means, explains Asif, that Katie encounters a great deal of difficulty in experiencing and processing her feelings, and in producing an appropriate emotional response to any given situation.

I can't help but think of how Katie's face is often oddly devoid of emotion when she's talking about difficult things. (James Joyce's famous line in *Dubliners*—Mr. Duffy "lived a little distance from his body"—seems to describe Katie's relationship to her emotions perfectly.)

Dr. Asif points out how Katie's left temporal lobe and prefrontal cortex lit up bright red in response to watching the burning tower. This area was also overstimulated when she thought about something sad.

But, Asif points out, moving back to the underactive right side of the brain, where her brain waves are moving too slowly, "Even though you felt panic, your body was *not* responding to these feelings of stress as it should."

Katie's motor cortex was also underactivated.

It is as if, says Dr. Asif, Katie's brain, at some point in her life, learned to shut off and stop receiving any input from her body—even as her prefrontal cortex and other areas of her brain were telling her she should be stressed and alert.

Dr. Asif looks at Katie as if he is reading the tea leaves of her suffering. He points to the right insular cortex and right parietal lobe again. "This slow wave activity, paradoxically, helps explain your panic events," he says. "It tells us that there is not a lot of energy in the emotional seat of the body. This suggests that when life feels overwhelming, or you feel anxiety, instead of sensing your feelings in your physical body, you shut down. You disassociate from your surroundings and go into denial. You go offline."

At first this may feel safer, he explains, "because you aren't feeling *anything*. But then, your anxious thoughts start spinning, taking over. You might feel yourself entering a kind of defensive posturing, blaming others, blaming yourself." This rumination and anxiety, "coupled with shutting down and not experiencing or expressing what you are feeling, can increase your anxiety to the point that you can't bear it," says Asif. "Life can seem so overwhelming that you simply do not have the ability to even move."

There is one other thing he wants to point out. When Katie closed her eyes during the dynamic brain scan, her brain waves showed something he often sees in individuals with depression. "There is no alpha peak here, even when your eyes were closed," he says.

"Alpha peak is a refreshing rhythm for your brain," Asif continues. "When it's missing when the eyes are closed, that tells us the brain can't rest, even in sleep. The brain has no respite. So you are very likely feeling physically exhausted much of the time, which will make you feel even more like you want to just shut off."

Katie nods slowly, following Dr. Asif's every word, as if she feels she's being seen for who she really is for the first time. His words convey the gift of validation.

"I think that this probably began when you were very young," Asif ventures. "We see this slow wave pattern in people with a history of trauma in early childhood. This is how a child's brain reacts. The brain goes into a traumatic freeze state, a detached mode, to keep from feeling the onslaught of emotions. There is a breach in how these circuits fire, and so we see deviations in these areas." He pauses. "We see this in many people."

Based on his clinical data, Asif ventures, he would even surmise that Katie experienced childhood adversity before she was two years of age.

At this, Katie begins to weep. "I'm sorry," she says. "I'm sorry." She starts to sob uncontrollably, as if all the emotion she's been holding back as she's listened has caught up with her.

Dr. Asif brings her a box of tissues and places his hand softly on her shoulder. We are all quiet. There is a well of compassion in this room.

Katie dabs her eyes again and again. "My parents used to fight all the time," she begins. "My dad would sometimes put his fist through the wall. He has a temper. There was a lot of volatility in my home."

"What would you do during those times?" Asif asks, his eyes settled on Katie's face.

"I'd try to make myself as small as possible. I'd try to tune it out." Sometimes, Katie recalls, "I'd go outside and make myself run as fast and as far as I could, back and forth in the yard. I'd tell myself that if I could just keep running and not stop, everything would be okay again."

Dr. Asif explains to Katie that influences in her childhood environment cannot be separated from biological factors. Experiences, relationships with caregivers, and trauma change the way in which an individual's brain becomes wired. From the moment we are born, our emotional responses and brain development are forged through our experiences with the external world. Every interaction with our caregivers, every sight, sound, and touch, influences new brain circuitry. And those changes in our brain in turn affect the way we respond to the world around us later in life, which continues to influence our experiences.

In Katie's case, says Asif, it's not surprising that even the smallest request life makes of her feels like too much. "Your right motor cortex is saying, 'Whoa, I don't want to do it,'" he explains.

A deep flicker of self-recognition crosses Katie's face. "When I look at this, and see my brain not responding, I feel overwhelmed," Katie confesses, gesturing with a limp hand toward the image of her brain twirling slowly in front of her in hues of blue and red and green on the computer monitor. "This *is* how I feel, like I cannot respond to life. Just yesterday I was so anxious about this trip I had to lie down, and I couldn't even take Andrew to his karate class." Tears stream down her face.

Dr. Asif sits next to her in his worn brown leather chair and leans forward, his elbows propped on his knees, fingers steepled in front of his face, as he listens, attentively. He seems to know, intuitively, that Katie is running on nothing but blind faith that there is anything left that can help her.

"It takes a lot of planning to arrange every detail so that even a trip to the grocery store won't trigger a panic attack," she goes on. "I have to shop in the midmorning when there aren't many people in the store. That way I can lie down for a few hours afterwards." Katie gives a self-deprecating half-laugh. "A few years ago, I taped a sign over our doorbell that said, in big, bold, black Sharpie, DO NOT RING OR KNOCK, JUST LEAVE PACKAGE, because the sound of the doorbell ringing or people knocking is too jarring for my nerves. I am always planning ahead to circumvent whatever might overwhelm me. It's *exhausting*. I'm a slave to my symptoms."

Katie brings the tissues that she's balled in her hands to her eyes again and again. "I feel as if you are looking into my soul. Everyone is always wondering what is wrong with me." She gestures again to the image of her brain. "Well, *that* is what is wrong with me, right there!"

"What I see here is very *hopeful*, Katie," Dr. Asif reassures her. "We can see what is happening here. And we know, based on thousands of patients, that we can address this."

"What about my panic attacks?" she asks. "Can you help me with those?" She wipes her eyes.

Hasan Asif smiles. "Panic is one of the easier things to treat," he replies. "I would even expect that by the end of your second week of treatment you will experience a twenty to thirty percent reduction in your symptoms." He pauses, then says, "Just give me two weeks to help pull you out of this."

In Katie's case, as in most cases, Asif says, he will combine transcranial magnetic stimulation with neurofeedback and vagal training. Vagal training, he says, will help Katie train her autonomic nervous system's response to psychological stressors. He and his team will show her how her breath and heart rate impact her brain scans and how her autonomic nervous system's response is affecting her perception of the world, and of herself.

"You will learn that *you* have the ability to control these areas and how you feel," Asif says.

But his most effective tool in helping her, he believes, will be transcranial magnetic stimulation. "We will put the pulse in areas on the right side of your brain, to help change the underlying neuronal firing power, attempting to activate the slow wave areas."

Katie wants to start treatment right away; she doesn't want to wait. Indeed, she's already made appointments for the rest of this week with the idea that if she decided not to opt for treatment, she could cancel them.

Countdown to Change

Day One

Katie is reclining in a large leather chair, this time in a small treatment room in the Bronxville office of Dr. Asif's Brain Wellness Center. Again, Asif has affixed a cap to Katie's head and inserted gel-glue into the holes that allow him to affix the electrodes to Katie's scalp. He then connects them to the qEEG and the low-resolution electromagnetic tomography machine. A few minutes later, he's running a live, dynamic brain scan—just as he did during her last visit. He's not treating Katie yet; he's showing her his treatment plan, and how, based on his findings, he will be using transcranial magnetic stimulation to help bring the areas of her brain that are not working properly back to healthy functioning again.

"First, we will be targeting this C4 region of the brain here, the sensory motor cortex, in the parietal lobe," he explains. "This area is responsible for processing touch and sensation in the body—or proprioception." He carefully adjusts the transcranial magnetic stimulator so that it will pulse exactly over just this area. Then he smiles at Katie. "Are you ready?"

I see a wave of anxiety wash over Katie's face. She seems caught up in a private inner dialogue that I can only dimly sense.

"Take a few deep breaths," Dr. Asif tells her, "and, if you can, relax your muscles." The respirator belt is affixed to Katie's abdomen and the sensors are attached to her fingers.

She is sweating. Katie asks me to help her take her sweater off, and I do, folding it on my lap under my clipboard. She lies back again. Her chest and arm muscles tighten, and her face is more ashen than usual.

Instead of commencing treatment, Dr. Asif gently moves the TMS machinery away. He points out to Katie that her brain scan is lighting up bright red in a pattern associated with a panic attack. He gestures to a second monitor and shows Katie that she is breathing a rapid seventeen shallow breaths a minute.

"Before we begin treating you with TMS, we are going to help you learn how to modulate your breathing, in order to help you bring down your body's physical stress response," he tells Katie. He believes, he explains, that "your panic events are directly related to dysregulation in your autonomic system."

For the next ten minutes, he shows her how to regulate her in-and-out breaths to match the rhythm of the respiration pacer she's also hooked up to—six to seven breath cycles a minute.

"You are always holding your breath, disengaging from what's going on around you, as if you are in freeze-panic mode," he says. A biometric chart on the computer in front of us compares Katie's breathing and heart rate in real time, as compared to the ideal, and alerts her when her breath is too shallow, or too fast.

As Katie begins to breathe in slower cycles, we can see, on her brain scan, that frontal lobe activity in areas that are too activated is beginning to slow down a bit.

This, says Asif, is what he calls *neurobreathing*: breath work to calm the neural firing patterns in the areas of Katie's brain that are overactive when she is anxious.

Once Katie has succeeded in modulating her breath to a slower in-and-out rhythm, Asif teaches her how to conduct a body scan, focusing, with her eyes closed, on each part of her body progressively, allowing herself to be present as she centers in on the bodily sensations in her head and neck, her shoulders, her chest, her arms, her hands, and so on down the length of her body, until she reaches her toes. When she finishes, he asks her to repeat this process.

It's only after all this prep, when Katie is calm, that Dr. Asif recommences transcranial magnetic stimulation.

He maneuvers the TMS stimulator over the same precise spot on her head again. There's a brief pulsing sound . . . *zzzz* . . . *zzzz* . . . *zzzz*, one, two, three. It lasts for just a few seconds.*

He checks in with Katie. "Doing okay?"

She is absolutely fine.

"I am going to do thirty sets," he says. There is another *bzzz* . . . *bzzz* . . . *bzzz* sequence.

The pulsing continues. Dr. Asif moves the TMS machine around, adjusting it and delivering pulses in three different spots. On the last

---

* According to Pascual-Leone, it's important to note that the electric current induced in TMS can be quite high, up to thousands of amperes. TMS safety guidelines stipulate that TMS intensity must be individually set for each patient by measuring that patient's brain's response to a single TMS pulse of variable intensity. This is known as determining the "motor threshold" and defines the stimulation intensity that will be used for a given patient. Also, he adds, the patient must wear earplugs to protect the ears from the clicking sound of the TMS device, which is actually very loud, even if it doesn't sound loud to the patient because it is so brief.

spot, instead of delivering many little sets of eight quick pulses, he delivers a series of longer taps, sixty times.

Throughout this process, he has Katie continue her neural breathing and body scanning.

And then that's that. The session is over.

"The rest of today, as you go through your day, I want you to focus on your senses," Dr. Asif tells Katie. "Notice how the light falls on the trees. Listen to sounds. This treatment is going to produce a lot of emotions in you. Try to stay in your body. Stay out of your head."

Dr. Asif will see her for another session tomorrow, and another the day after. Then she will go home to be with her children and come back again early next week for another week of treatment. Initially, Dr. Asif says, he likes to treat a patient who is coming in from out of town at least three times a week, in order to help the brain receive the message that these changes are the new way that the brain will be in the world, so that the brain can start to reregulate itself.*

As we leave the office, I ask Katie, "How do you feel?"

"I'm tired, but I feel pretty good!" she says. She's headed back to her hotel to rest. She also wants to change her travel plans. She's currently scheduled to leave on a train that won't get her home until midnight. The travel is enervating for her, so she wants to go home a day later so she can rest a bit before she sees her kids.

Day Two

"Did you have any dreams?" Dr. Asif asks Katie the next day, after she's been hooked up to all of the equipment. They sit facing each other. Asif leans forward, his hands resting between his knees. The tips of his thumbs and fingers touch each other lightly, forming a shape that looks like a yogi heart mudra.

She nods. "I dreamt I was floating over my kids, hovering over them

---

* FDA-approved guidelines are for five days of treatment for up to six weeks in a row.

as they played, and in my dream I felt this terrible heavy sadness, like a cloud, all around me."

"What was the sadness?"

"This feeling of grief that I'd missed out on so much with my kids all these years. And then in my dream I started to feel nauseated, I had all these butterflies in my stomach, I started to panic—and my panic woke me up." But, Katie says, "I did the breathing technique and I was able to calm down. I was able to go back to sleep."

"How do you feel now?"

"Tired. Yesterday I had to reorganize my travel plans to stay another day, and it's hard for me to make lots of small decisions about trains and babysitters and all of that. It's really stressful."

"I think I notice your eyes misting over," Dr. Asif says. "Are you aware of feeling sad or anxious?"

She nods. "I was afraid if I stayed another day it would be too hard on my kids," she says. "But I was afraid if I didn't stay another day and rest up before I travel home I wouldn't feel well. I was caught between wanting to pace myself and making my family happy." Katie pauses. "I guess it's the worry I always have, of not feeling good enough no matter what I do."

"That is one of your primary fears, from early in your life," Asif says.

Katie's eyes are wet. "I'm just feeling the loss of all that I've missed." A tear trickles down her cheek.

As Katie continues to talk, Dr. Asif gets up, telling her he is going to do a TMS treatment. "We are going to treat the area that has to do with decision making," he says.

Dr. Asif follows up TMS treatment with a neurofeedback treatment. "We want to solidify the new brain patterns we're hoping to establish," he says.

"How will I know if this is helping me?" Katie asks.

"Often, other people will notice changes in you first. They'll see that you are responding differently to the world before you do," he says. "As your brain begins to learn new patterns, you will start to see changes in how you live your daily life."

Day Ten

A little over a week later, Katie and I meet again in Asif's waiting room. "How are you feeling?" I ask, pen in hand.

"I think the best way I'd describe it is to say that I feel a little more *alive*," she says. "A little less *dead* inside." She gives a small, happy laugh. "Things are just . . . feeling easier for me!"

I'm startled. It has been only eight days since I saw her last; I did not sit in on her second week of treatment, and she has been home with her family since then. There is so much more affect in her voice.

"It is hard to put into words," she goes on. "I feel . . . *lighter*."

"In what way?" I ask.

"For one thing, I no longer have to take a two-hour nap every day, and that, for me, is life changing."

She tells me about yesterday's neurotherapy session with Dr. Asif. "He explained to me that he would be working on areas of my brain related to early attachment," she says. "He asked me about difficult events in my childhood. I talked about how lonely and isolated I felt by the time I was twelve or thirteen—how ugly, how unattractive—just not good enough no matter what I did. And I could see my brain light up in areas that he explained were related to attachment, panic, and shutting down."

After treatment with transcranial magnetic stimulation, those areas changed, Katie says. And she felt something shift inside. "For the first time in my life, I feel like something is changing in how I view *myself*. I don't feel so overwhelmed by self-loathing, maybe?"

Half an hour later, Katie is hooked up to the neurotherapy equipment, waiting patiently for Dr. Asif. Katie keeps looking at her phone to check the time; she has a train to catch in just a few hours. A full thirty minutes after her appointment time, Dr. Asif walks in without explanation for his tardiness—though he will later explain that he was simply running late with another patient. "How are you feeling, Katie?"

"I feel good!" Katie tells him, resolutely, but with a tinge of anger in her voice.

"When you say 'good,' tell me what that means for you?"

"Well, I feel good but I also feel angry," Katie says, her voice tight.

"When you say 'angry,' what do you mean?"

"I feel invalidated," she confesses to him. "It's a lot for me to get here, with my kids at home and my parents and babysitter managing every-thing for me. I've told you travel makes me anxious. It just makes me angry that even knowing what I'm doing to be here, and make this treat-ment happen, I'm just sitting here, waiting for you."

"Is that sense of invalidation an old feeling?" Dr. Asif asks her.

That question appears to hit Katie unexpectedly hard. She breaks into tears. "Yes," she confesses to Dr. Asif. "Yes, my mom would always tell me, 'Katie, you are too emotional! Life is too hard for you! You have to get a thicker skin!'"

"It was a lack of validation, that whatever you were feeling it was not okay?" he probes.

"There was never a time when my parents encouraged any emotion. Feelings didn't matter. It was always about whether I was doing well enough. Was I going to get into a good college? Was I doing and achiev-ing enough?"

As they converse, Dr. Asif begins treating her with TMS. By doing the dynamic live brain scan simultaneously with TMS, he explains, "We can make certain that we are on the right track."

Dr. Asif begins to point out certain patterns in her brain as they talk. "When you are in quiet, pleaser mode, holding in your anger, we see patterns in the right side of your brain which show us the trouble you are having feeling your emotions," he says. "But when we put the TMS pulse on this freeze zone—and you begin to allow your emotions— notice how these areas begin to come back into homeostasis." He points to her left temporal lobe. "This whole area was dancing around, and now that we've directly stimulated the cortical neurons, see how they are in a new firing pattern? Your brain really straightens out with the pulsing." He paused and then added, "Our perceptions are *based* on our neuronal firing pattern. As we help that firing pattern to change, you will feel many things shift."

He points out another major change he sees. "Your brain waves are now showing alpha peaks right after stimulation. This is very good news. It also tells us we are on the right track."

"How are you feeling in your body now?" Dr. Asif asks at the end of today's treatment.

"A sense of physical and emotional peace, I guess," Katie says. "Relief?"

"The real effect is the difference you'll begin to see when you are going through a stressful event and you are able to respond to the world around you differently," he tells her. "The major changes will come about not when you are in treatment, but over the coming weeks. It will become more rewarding—instead of frightening—to pay attention to your body's emotional state, to be in a state of introspection. And then, as you notice how you are feeling, you will be able to speak up, to respond differently, more assertively, from your own felt sense, instead of detaching and going into panic mode."

"Has anyone around you noticed a change in you?" I ask, as we leave Dr. Asif's office together.

"This morning, when I called my mom to check in on my kids, she said, 'Katie, you sound *different*, your voice sounds happier, less stressed and tired,'" she says. "And I told her, 'I feel good!' My mom was quiet and then she said, 'I can't think of a time when I've heard you say those three words.' She was really happy for me.

"It's Andrew's birthday this weekend," Katie continues. "I'm excited to celebrate my son's birthday! I've never felt this way about planning an event. Two weeks ago, I would have been really stressed about it, just gutting through it to make sure that the kids have a good time. It would have been agonizing for me. But now, I feel, I don't know, happy to be able to *be* there, mentally, with my kids."

Day Seventeen

At the end of the third week of treatment, Katie and I meet again in New York City. She's just had her seventh, eighth, and ninth TMS treatments.

We are sitting in a Starbucks on Forty-Second Street, which is noisy and crowded. We're seated near the bathroom door, and people are lined up not two feet from the edge of our table. A mother is holding her crying baby in her arms.

"You're sure you're okay sitting here?" I ask her, wondering how she is handling being in such close proximity to the chaos that comes with human togetherness in public spaces.

"Yes!" Katie says. "I am fine! It's . . . amazing to feel that yes, I can sit here, I can do this."

"And yet when we first met . . ." I start to say.

"I know! I know!" she says. "I couldn't tolerate coffee shops. I would have had to get out of here. Things are just . . . easier for me now. Even being here in the city, I'm okay navigating the streets, with the noise, the traffic. The last time I was in the city I felt terrified every minute." Katie's voice has . . . color. There is a lilt in it I have not heard before.

"I'm able to have a really full day for the first time in my life," Katie continues. The corners of her mouth and eyes crinkle, turning up as she talks. I remember how, when we first met, it was as if her facial muscles had only two settings: out of commission and an over-the-top smile.

"Usually by noon I'm feeling I have to lie down." She looks at her watch. "Now it's two o'clock and I'm still feeling good.

"There is something else I've noticed that's different," she adds. "A few times in the last few days I've found myself singing. I have never been into music, it's always been too stimulating for me. I've always had to have quiet in the car, at home. If the kids wanted to listen to music they had to wear headphones. But now . . . well, I'm singing in the shower. In the car I'm singing along to the kids' music with them."

She's somber for a moment. "I realize how much I've missed of their lives, and how much they've missed out on because of my limitations. So much of the time, instead of *being* with them, my mind has been a magnet for horrific worries about something that might happen to them. I remember just three or four weeks ago, I was thinking about whether we should have hot dogs at Andrew's birthday party. And then I was trying to remember how to do the Heimlich maneuver. And then I called the babysitter and my mom and told them we had to practice it before his party. Instead of being with my kids I've been caught up in one stressful, panicked thought after another. And realizing that brings up grief too.

"I've been followed by this ever-present, amorphous sadness and sense of panic and dread for as long as I can remember," she continues. "I've always known I didn't feel the joy that other people felt. And then there have been those moments that have been so bad, when I've been so deep in the well, that I've felt that being dead would be a relief, and maybe it would even be a favor to others if I weren't here."

Feeling such a profound shift in her emotional well-being, she explains, as she twirls a mint tea bag in her mug, makes her feel a residual anger toward "a society that assumes depression and anxiety are a choice. Over the past twenty years so many people have told me 'You worry too much!' or 'You can get over this if you try hard enough!' My pain made them feel uncomfortable. Living in our society we all buy into that myth, that somehow depression is optional."

And another thing that working with Dr. Asif has done for her, she says, pulling the tea bag out and taking a sip of tea, "is to help remove that sense of shame I've carried. I can see that what's happening in my brain *is* real. I can see how my brain is qualitatively different from other brains. And I can see, as we begin to change how my brain fires up, I feel different in the world.

"For me, this is nothing short of a miracle," she says. "And I am only a third of the way through my treatment."

Day Twenty-One

Katie calls me after she's returned home from having had four more treatments—thirteen in all. She wants me to know, she says, that she has been able to cut the doses of two of her medications in half. She has been checking in with her psychiatrist in Virginia as well as Dr. Asif, and the two psychiatrists agree that she no longer needs to be on such high doses.

I can hear her doorbell ringing in the background.

There is something else she wants to tell me, she says, talking quickly. "I took the sign off the doorbell!" she says. "The one that said DO NOT RING OR KNOCK! I just didn't need it anymore." She pauses. "I feel almost as if the woman who needed to live in a world in which the doorbell never rang was a different woman, a version of me, yes, but not *me*. I feel like I'm getting back to being *myself*.

"I have to go!" she says, before I can interject a question. "I have a group of boys and their moms coming over for a playdate with Andrew this afternoon and they're here!"

I'm stunned. In the background, before she hangs up, I hear Katie's doorbell ringing over and over again, and her son calling, "Mom! Mom! They're here!" and it sounds to me like the chaotic and happy music of family life.

Later, by phone, I talk with Dr. Asif and tell him that I am astonished by the change in Katie. I tell him that I know that not every patient will respond to TMS as well as Katie (some patients in clinical trials don't respond at all). And, I offer, I'm also aware that there are serious cautions regarding who can have TMS—people with pacemakers, for instance, can't receive treatment because the magnetic charge can turn off their pacemakers.* And TMS has not been well studied in patients with bipolar disorder; there are concerns that in some patients with bipolar depression, TMS might worsen their symptoms.

And then there's this on my mind: As a science journalist, I feel somewhat concerned as to whether these changes will be long-lasting. Could part of Katie's response possibly be placebo? It is hard to fathom that a psyche such as Katie's, which has been in such inconsolable pain for so long, could be so quickly mended.

Asif tells me that he believes that this is what healing can look like when we bring together the safest, most effective tools that already exist

---

* There are also scattered reports of patients experiencing seizures during treatment. As with all medical treatments, patients interested in TMS should, of course, consider their own medical health history, seek advice from their primary care physician, and be mindful of how experienced their provider is in providing TMS to his or her patients.

for helping the brain come back to a state of homeostasis. When microglia back off, neurons begin to repopulate and fire off in healthy, positive ways again.

Still, despite both study results and clinical successes showing how TMS benefits some patients with treatment-resistant depression and anxiety, the field of psychiatry has been reluctant to add it to treatment plans, continuing to rely instead almost solely on a pharmaceutical approach. Expense is also an issue. Neurotherapy equipment costs tens of thousands of dollars—money that most psychiatrists would not normally invest. And it is not inexpensive for the patient. After Katie's insurance reimburses her for thirty sessions, she will still likely face several thousand dollars in copays (though this is similar to the amount she would pay had she sat in most psychiatrists' offices for twenty or thirty sessions).

Asif hopes, he says, that one day, with better insurance coverage, neurotherapy will become affordable for vulnerable groups everywhere: teens in high school, for instance, "so that we can begin to train their brains to fire in healthier ways before psychiatric disorders set in," he conjectures. "And I'd like to see this in community centers for the elderly, to help them with anxiety and memory loss."

A New Era in Medicine

All of this leads us toward a larger quest: If microglia manipulate our health and our brain, what are other promising ways that we can safely manipulate and reboot microglial cells that have gone off kilter in order to help neurons and synapses regrow and the brain to heal?

Today, in both clinical research labs and therapeutic settings, scientists are uncovering extraordinary approaches to help calm overreactive microglia so that they behave as nature intended: as the angels of the brain, rather than as blind assassins.

The wonderfully neuroplastic brain is exquisitely responsive to intervention, once we figure out how to intervene.

TEN

Untangling Alzheimer's

I T'S AN UNSEASONABLY HOT DAY IN THE MIDDLE OF MAY IN CAM-bridge, Massachusetts, with the temperature topping 95 degrees. The heat is slowing everything down. The leaves on the ginkgo trees that partially shade the old brick sidewalks between the Massachusetts Institute of Technology and Harvard hang limp. Pedestrians duck into doorways to dab at their damp brows. The whole city seems to be running in slo-mo.

But inside the cool gray glass-walled tower that houses MIT's Picower Institute for Learning and Memory, Li-Huei Tsai, Ph.D., a cognitive neuroscientist and director of Picower, is talking animatedly as she explains her 2016 discovery that demonstrates how microglia can be hacked and marshaled to help reverse the biological ravages of Alzheimer's disease.

The idea that flailing microglia can be rebooted to treat Alzheimer's patients may sound like Star Trek–era medicine, but Tsai and her team have already entered human clinical trials, utilizing a brain hacking method known as GENUS—short for gamma entrainment using sensory stimuli, and more casually known as gamma light flicker therapy—

to reprogram microglial cells so that they stimulate synapse regrowth and clear away plaques and tangles.*

Tsai tells me about her largely accidental discovery as we sit at the oval conference table that dominates her office. Between us sits a foot-long clear plastic box, which looks like a large Tupperware container. Inside the box, a tiny computer circuit board and processing unit are secured with black duct tape to multiple cords, which connect to a short black plastic wand that extends out of a hole cut into the box's side. Several small LED lights are attached to the end of the wand.

This hand-engineered contraption appears so simple it looks more like a middle school science project put together at the kitchen table than a scientist's breakthrough discovery of a lifetime. But the story behind its precision high-tech engineering and Tsai's paradigm-shifting discovery is already legendary among neuroscientists.

In 2015, Tsai, who was then in her midfifties, was thinking about several genes that had just been found to increase the likelihood of Alzheimer's disease, and she was feeling frustrated. Many dozens of gene mutations appeared to play a role in each step of the disease, and she feared it was "going to take forever," she says, to translate that science into meaningful interventions.

So Tsai began to think about Alzheimer's from a computer engineering perspective instead. "The brain really functions as an information processing unit," she tells me, her fingers folding around the front of her chair's armrests as she talks. "The brain is making billions of rapid connections within milliseconds that constantly send forth new instructions. And this complex processing is guided by rhythmic brain waves, which organize neurons into ensembles in different regions of the brain so that they can do their separate jobs." It is, says Tsai, "a lot like a small orches-

---

* *Plaques* are deposits of proteins called beta amyloid; these proteins eventually form amyloid plaques, which clump together around dead nerve cells and disrupt cell function. *Tangles* are abnormal accumulations of a protein called tau that collect inside neurons; this tau protein sticks together and twists into tangles that harm the synaptic communication between neurons, resulting in the symptoms of Alzheimer's disease.

tra with different parts, all playing different notes—but at the same time these groups have to work in tandem, under one conductor, to produce harmonious music."

And the master conductor that *all* regions of the brain respond to is electrical signaling.

Tsai and her team wondered: What might be the first signs that the brain's electrical signaling was starting to go awry very early on in Alzheimer's, long before symptoms appeared?

One way to measure this, Tsai realized, would be by looking at brain wave rhythms. And it turned out that, indeed, one brain wave in particular, known as the gamma rhythm, was "really compromised" in early Alzheimer's. Gamma waves are crucial to higher brain wave activity, attention, perception, and memory. Scientists at the University of California, San Francisco, had already demonstrated that in Alzheimer's, disruptions in gamma brain wave activity were notably evident in areas of the brain associated with complex thought, especially in the hippocampus and the prefrontal cortex, regions crucial to memory and attention.

That gave Tsai an idea. "I wondered, what if we could increase gamma power and electrical signaling back to what it should be in the hippocampus—to the way it was before Alzheimer's?"

At the time, Tsai was hoping that by altering the brain's signals, she could change the activity of neurons. "Microglia were not really on my radar at all." She laughs. "I was hoping that by manipulating gamma waves, I might be able to restore the function of neurons within the hippocampus."

Tsai had already found, in animal studies, that delivering light pulses at 40 hertz directly into the brain's hippocampus, through a technique known as *optogenetics*, stimulated neurons. But this technique had major drawbacks. "Optogenetics is quite invasive," she says. "We have to insert fiber optic wires directly into the hippocampus and shine the laser light inside the brain. Then we use the laser light to control the electrical signaling and activity of neurons." It certainly wasn't an approach that could be translated into everyday treatments for patients with Alzheimer's within the foreseeable future.

Tsai and her team of neuroengineers set out to create a computer program that would allow them to deliver targeted pulses of light to the outside of the brain, noninvasively, in animals. They programmed their computer software so that a strip of LED lights on a handheld light wand would flicker at the frequency of gamma wave oscillations. Then, using this new LED-based light-flicker apparatus, they began to treat mice with a version of early-stage Alzheimer's with what they called gamma light flicker therapy.*

"We began delivering the gamma light flicker therapy at forty hertz to mice," Tsai explains, excitedly, her smooth oval face lighting up as she recounts these initial moments of discovery. "We delivered this gamma therapy for one hour." Tsai occasionally looks at the plastic box that sits between us as she talks, as if she sometimes still finds herself amazed by what she's accomplished using this simple apparatus.

The immediate results of Tsai's experiment were so astonishing that Tsai doubted what she was seeing happen within the mouse brain. "After one hour, there was a forty to fifty percent reduction in the amount of beta-amyloid proteins in the part of the brain we were looking at, the visual cortex, which processes vision."

Tsai discovered something else entirely unexpected. "It turned out that gamma light flicker therapy was not only affecting the activity of neurons, it was also altering the activity of *microglia*," Tsai says. "By delivering gamma waves at that specific frequency, we had somehow caused microglia to do their job again. They started to clear away amyloid proteins!"

Microglia didn't just become good housekeepers. They became an enormously efficient emergency clean-up crew, mopping up the amyloid proteins. And they also burst apart, digested, and removed already existing beta-amyloid proteins. "In just one hour, microglia cleared the brain of *half* of its toxic amyloid proteins without damaging brain tissue." Again, in this experiment, the mice were in the early stages of Alzheimer's.

---

* These young mice with a version of Alzheimer's had developed amyloid proteins, but the proteins had not yet formed into amyloid plaques, as they do in older mice.

In fact, not only were microglia clearing plaques, once they had broken down beta-amyloid protein chunks into tiny little bits, those bits became completely harmless debris. They were now small enough to drain away through the brain's lymphatic pathways—the same lymphatic immune vessels that Jony Kipnis had recently discovered ran through the meningeal spaces of the brain and connected to the body's lymphatic immune pathways.

Tsai's discovery was "a total accident," she admits, leaning back in her chair and tossing her hands up in the air with a flash of glee that I suspect is rare for this neuroscientist who manages a team of thirty at one of the world's most renowned research institutions. "This is the beauty of research," she says. "You just never know where the science is going to take you."

Still, when Tsai first saw the results of her experiment, her only thought was *Will we ever be able to replicate this?* She lay awake for three nights straight, she confesses. "It was hard to wrap my head around the idea that what we were seeing was true. I kept thinking, *What information can I gather that will convince me that what I am seeing is really happening?* And I thought, *If I can see precisely what molecular events are altered, and explain these changes to* myself, *I'll feel more confident about our findings.*"

So Tsai and her team stained tissue in the mouse brain's hippocampus so that they would be able—while using their original, more invasive brain-stimulating technique, optogenetics—to see microglia glowing and observe them more closely inside the brain. Then they turned on the laser light stimulation device and watched and waited. What they observed was mind-blowing.

Before the mice were treated, "the microglia just sat there, not doing much. They were just secreting neurotoxic cytokines," says Tsai. These microglia appeared small and sickly. "But after just one hour of optogenetics the microglia puffed up and became twice as large—they got all healthy-looking and they got busy! After an hour, ninety percent of microglia were starting to clear up amyloid proteins. In just one hour we were able to bring them back to do their job!"

But there were two big catches. First, after twenty-four hours, the amyloid proteins returned to their original levels. Second, optogenetics requires creating a physical hole through the skull.

So Tsai and her team returned to using their noninvasive GENUS technique to see whether a longer course of treatment of gamma light flicker therapy—delivered safely from outside the skull—could more permanently reduce amyloid plaques. This time, they treated older mice—mice that had already developed the clumps of amyloid plaques associated with later-stage Alzheimer's—for an hour a day for seven days.

This time, they saw something entirely new. Longer-term treatment with gamma light flicker therapy, says Tsai, led to a reduction of the clusters of amyloid plaques associated with more advanced Alzheimer's.

Delivering repeated treatments of pulses of gamma light flicker therapy caused microglia to change from inflammatory chaos agents, secreting toxic chemicals that sickened other brain cells, into the proactive guardians of the brain that nature intended them to be.

After treatment, microglia became remarkably fluid. They soothed neurons, bathed them in neuroprotective, anti-inflammatory factors, repaired vulnerable synapses, and stimulated new neurons to grow, repopulating neurons in the hippocampus. And they did all this while continuing to scour the brain for plaques and tangles and clear them away. This time, the results were no longer temporary but long-lasting—for up to a week.

The brain began to heal itself.

"I feel I live in a dream—it's almost magical, surreal," confesses Tsai. "I feel as if one day I'll wake up and realize, no, this can't be true." She rests both of her palms on the table in front of her and takes in a breath. "But it *is* true," she says, her dark brown eyes sparkling. "And now we are looking to see if this can work in the human brain too."

Tsai published her findings first in the journal *Nature* in 2016, and in a follow-up study in *Cell* in 2019. Meanwhile, other researchers were also finding that delivering specific, noninvasive doses of ultrasound waves influenced microglia to clear Alzheimer's plaques in animal models. Scientists at the Queensland Brain Institute at Australia's University of Queensland discovered that repeated ultrasound treatment reduced amyloid plaques in the brains of mice by 75 percent—and treated mice showed improved performance on memory tasks, including recognizing objects and running through a maze.

Tsai's and other researchers' findings have helped to redirect the field of Alzheimer's disease research to treat it as a disease of the brain's immune system—a disease of microglia. In 2018, Tsai published a paper sharing her team's instructions on how to construct a noninvasive LED-based light-flicker apparatus—or GENUS—in hopes that other labs can start to utilize gamma light flicker therapy in their efforts to better understand how to treat not only Alzheimer's but possibly disorders of the brain such as autism and schizophrenia.

"We know that microglia should be there to get rid of the plaques and tangles that build up in the brain, but for some reason, in Alzheimer's they stop helping," says Tsai. "Instead, they make the brain more inflamed."

So what is it that makes good microglia turn bad?

The Microglia-Gene-Alzheimer's Connection

Remember that in 2016, Margaret McCarthy's University of Maryland lab had demonstrated that once microglia are triggered to enter a protracted inflammatory state, this causes changes to the genes that oversee how microglia will behave over the long term, causing runaway neuroinflammation. And the more neuroinflammation there is, the more likely microglia are to take down synapses, spew out inflammatory chemicals, and stop clearing amyloid plaques and tangles.

When microglia morph from being the good doctors of the brain and instead start destroying brain synapses, they do so because the genes that should tell microglia to behave properly have been triggered by something (some uniquely personal combination of genetic predisposition and influences in an individual's environment, including stressors, trauma, infections, injury, toxins) to undergo what are called *epigenetic shifts*—alterations in what we might think of as their operating code.

These epigenetic shifts cause genes to erroneously send microglia out to plunder synapses, spit out toxic chemicals, and let debris build up in disruptive trash piles.

Alzheimer's researchers have discovered that a number of genes significantly increase the risk of developing Alzheimer's disease, and many of these genes, thus far, are specific to microglia. For instance, when there are mutations in a receptor gene known as TREM2—a gene that exists only within microglial immune cells—microglia have a diminished capacity to defend the brain against amyloid plaque. Indeed, people with variants of this mutation have a threefold greater risk of developing Alzheimer's.* In normal circumstances, TREM2 receptors aid microglia in protecting the brain against amyloid plaques by helping microglia form big armlike protrusions that extend out protectively to envelop, contain, and curtail early-stage plaques. But in vulnerable brains, TREM2 receptors suddenly lose their ability to help microglia combat plaque. Instead, this gene undergoes shifts in how its regulatory capabilities are expressed, so that microglia, conversely, produce the neurotoxins and pro-inflammatory cytokines that increase plaque production.†

But knowing this doesn't help scientists as much as we might think. As Tsai has pointed out, there are potentially hundreds of microglia-specific genes like TREM2 that are involved in turning on or off microglia's good and bad properties. And that means that "finding a way to affect one gene or even several genes is unlikely to be powerful enough to restore brain function," says Tsai.

Microglia, on the other hand, she emphasizes, "are like one giant light switch" through which many gene signals pass. "And we believe that by using gamma light flicker therapy to flip that switch, we can turn on microglia to do their job as they were intended to do."

* TREM2 is an acronym for the gene's weighty scientific name: "triggering receptor expressed on myeloid cells 2 protein."

† TREM2 is just one example of many microglia-specific regulatory genes that play a role in whether microglia will behave in a good way or a bad way, and whether and when any given individual will experience Alzheimer's. And all of these genes are "expressed" in microglia.

The Race to Find Biomarkers for Alzheimer's

Six months before Tsai published her groundbreaking study, Beth Stevens had published her seminal study showing that microglia nibbled away at complement-tagged synapses very early on in Alzheimer's disease. In fact, Stevens had found that synapse loss was an early event in the Alzheimer's disease cascade. Microglia were congregating and destroying healthy synapses in the brain long before plaque formation began, and decades before symptoms set in—at least in animal models. The correlation between synapses' disappearance and later development of the first signs of cognitive decline turned out to be even stronger than the correlation between cognitive dysfunction and the presence of plaques and tangles in the brain.

It's like a chain reaction. Microglia are triggered to eat synapses. Plaques and tangles build up in chunks. These iceberg-like proteins can't drain out of the brain as they should. Triggered by the presence of beta-amyloid plaque deposits and tangles congregating in the brain, microglia become even more inflammatory and start to engulf more synapses and damage neurons and trigger other glial cells, such as astrocytes, to secrete neurotoxic factors. And a patient's cognitive and mental state begins to deteriorate.

Later, over Skype, I ask Beth Stevens why this revelation—that synapse loss precedes the formation of plaques and tangles—is so important in the prevention and treatment of Alzheimer's (including with gamma light flicker therapy). Stevens is in her usual caffeine-fueled hurry—tomorrow she's taking her entire team, along with her husband and two daughters, on a "labcation," she tells me. "Everyone has been working around the clock here," she says. "I've rented a bus and a driver and we're going to head out to the Cape for the day to windsurf and kayak and sail. We need this!" (Not for the first time, I think that Stevens's zest for adventure, her moppish curls, and her scientific enthusiasm bring to mind the PBS television character Ms. Frizzle—the animated science-whiz teacher who whisks busloads of young people away on wacky and unexpected expeditions.)

"Plaques and tangles are clearly bad for the brain in many ways," Stevens tells me. "And plaques form early on in the disease. But the cognitive decline that we commonly think of as the hallmark of Alzheimer's disease is associated with synapse loss. We have so much evidence that synapse loss happens really early in the hippocampus, long before patients get clinical symptoms, that we know it is an early event. And we believe if you can catch synapse loss when it happens at the very beginning, it becomes a very different story." We also know, she continues, that "once runaway inflammation sets in, it is hard to reverse, so that means we are probably going to need to find a way to intervene with that process very early in the disease." Because if you think about it, she adds, "the reason that clinical trials for new drugs in Alzheimer's have failed so far is that they aren't starting to intervene early enough." As with most types of cancer, if you start treatment once it has already metastasized, in the later stages of disease, it's too late. And that means, stresses Stevens—returning to an area of research she's long been passionate about—"we *must* develop early biomarkers to detect when pruning *first* happens, before inflammation sets in. If we can do that, we have the most hope of preventing disease, because then we can start to treat patients early.

"We know that a large number of the risk genes for Alzheimer's are specific to microglia, which gives us more evidence that microglia are causal in the development of Alzheimer's." But, she continues, echoing Li-Huei Tsai's earlier thoughts, "We also know genes are not going to be our only answer." Rather, genes will be part of the answer in helping researchers to create better early biomarkers for detecting—and then treating and even preventing—Alzheimer's.

Imagine a time, Stevens poses, when clinicians can see exactly when some of the genes that regulate microglial behavior stop sending good signals and start to send nefarious ones instead. A time when we not only know exactly when good microglia become bad microglia, but we can also step in and reverse that process, even before they begin snipping off synapses.

Today, Stevens's colleagues at the Broad Institute are searching for the very first signs that microglia-specific genes are starting to emit destructive signals. She describes several new emerging technologies that are helping researchers to discern these signs. One of the ways they're able to do this is through another modern feat of engineering—experimenting with a relatively new invention known as *organoids,* or what Stevens describes as "mini brains engineered out of human stem cells that can be programmed so that they become different neurons and glial cells, including microglia." In these models of the brain, scientists can manipulate microglia with different genetic backgrounds (including variants of the TREM2 gene) and observe how and when microglia misbehave in different contexts.

"We are trying to go back to the very beginning to see how synapse loss starts, why some synapses are vulnerable, and better figure out what makes microglia transition from a healthy state to an activated state," Stevens explains. "We now appreciate that microglia have many activation states, some that are beneficial and some that are detrimental. For example, when microglia begin to change, they send out new molecular signals between cells that show that they are becoming neuroinflammatory. And if we can identify what those signals are, and find ways of assigning molecular markers to their different states, then we can see, hey, this microglia is getting ready to be an overpruner. And this microglia is getting ready to put out inflammatory signals. And this one is eating up amyloid."

Once scientists discern and identify these different molecular fingerprints, they will be able to know what microglia are about to do before they do it.

Another way that scientists are interrogating how microglia differ in varied states of activation is through a new technology called *Drop-seq* (short for droplet sequencing) in which researchers, says Stevens, "take microglial cells and put them in individual droplets with a kind of barcode that allows us to read out their DNA."* In this way, researchers can

---

* Using new Drop-seq and single-cell RNA sequencing methods in order to profile gene expression changes in individual microglial cells is, says Stevens, the newest and most transformative technique scientists have—and it is enabling them to characterize microglia in mouse and postmortem human brain samples for the first time, in ways that researchers hope will have profound implications for understanding and intervening in disease.

see what genes microglia are expressing compared to their neighbors, collect and profile individual microglia, and create a huge database that will help them to label and categorize different activation states of microglia. It's a little like the way in which we use a supermarket barcode to tell us where a product came from, what it contains, and when it will go bad.

With these new technologies, says Stevens, "We can hopefully identify genes or pathways in microglia that are causing them to misbehave and damage neurons. Those signals are a proxy for telling us that something is going wrong at a very basic level."

Imagine, says Stevens, that one day "researchers can start to recognize specific molecular fingerprints"—these barcodes—as microglia start to morph from angels to assassins "by actually *scanning* for them in the human brain."*

Sometime in the next decade, Stevens believes, scientists will be able to determine the precise molecular fingerprint of microglia that are poised to eat synapses. Once researchers know what those markers are, they can safely introduce radiotracer dyes in human brain tissue so that those distinct biomarkers—these genetic barcodes—will light up and glow in vulnerable circuits while a patient undergoes a PET scan. (This is similar to the way in which we currently take images of areas of the body via CAT scans to look for tumors, which are then diagnosed as cancerous or benign through biopsy.)†

---

* Already, a team of researchers at Yale has found a way to crudely quantify synapse density and loss in living people, by injecting radiotracers that cause specific brain proteins to light up and using PET scans to image the brain. This technology lacks the fine-grained specificity that Stevens envisions one day becoming available using microglial molecular biomarkers—but it's a step in that direction.

† There are several factors that make treating Alzheimer's even more complex than targeting cancer. In treating cancer, physicians can utilize surgical approaches, removing tumors and nearby tissue, as well as deliver immunotherapies to the site of the cancer itself. But it is not possible to surgically remove Alzheimer's plaques and tangles, and it is also far harder to deliver biologics into the brain, due to the blood-brain barrier, than it is to deliver cancer drugs into the bloodstream. This means that Tsai's and other researchers' discoveries of potential noninvasive interventions are all the more important and exciting.

"If we can detect a glia-neuron circuit misbehaving, a synapse starting to go, and know precisely when and where it's happening, just by putting a scanner over the brain, or if we can home in on just a few glial cells impacting vulnerable circuits, then we may be able to intervene decades before symptoms begin," says Stevens. "And that could be a game changer."

Of course, injecting radiotracer dye into people's brains and having them undergo PET scans, which use radiation to image the brain, hardly qualifies as routine testing—especially if individuals haven't shown symptoms of Alzheimer's. And yet we know that early changes occur in the brain decades before symptoms of Alzheimer's emerge. For this reason, radiotracer dye PET scans may prove a wise or viable future screening option only for individuals who possess abnormal TREM2 and other key genes that dramatically increase their risk for Alzheimer's, and who also have a strong family history of the disease.

Meanwhile, as scientists work to identify factors that microglia secrete as they begin to shift in form and function, they are also working on ways to recognize signatures of these factors in human blood samples.*

"Our goal is to one day develop predictive blood and fluid biomarkers that will give us enough good information that we can predict psychiatric and cognitive disorders before they happen," says Stevens. Blood tests for synapse loss are still a long way from being available in your doctor's office, but the field is urgently headed in that direction.

Let's imagine all of this in a potential real-life scenario.

Imagine a woman in midlife, like Katie, whose grandmother died of early-onset Alzheimer's. Now imagine that she has tested positive for several genetic mutations, such as the TREM2 and ApoE4 gene mutations, which are known to increase an individual's chance of developing early-onset Alzheimer's. Just as a woman with a family history of breast cancer

---

* We'll learn more about how scientists are working to develop blood tests that indicate microglia are changing their activity in the brain in chapter 13, "In Search of a Fire Extinguisher for the Brain."

might go in for more regular mammogram screenings, Katie might go in for microglial PET scans every year. Let's say she's forty-five and her yearly microglial PET scan shows that a certain small area of her brain is lighting up, indicating that regulatory genes expressed in microglia are starting to change in an ominous way that's been linked to Alzheimer's. Synapses in a specific area of the hippocampus are going to go down.

Katie's radiologist gives the report to her doctor, who then sends her for prophylactic gamma light flicker therapy. She will very likely also have Katie work with a behavioral therapist to adopt specific dietary and exercise programs that have also been shown to help create neurogenesis. Knowing too that less-active synapses are more likely to be tagged by complement and pruned away, Katie might learn a new language, start doing Sudoku, or take up knitting—or all three. She'll probably work hard to avoid environmental toxins, pathogens, infections, and stressors, which have been shown to contribute to triggering overtaxed microglial immune cells to become hyperactive. By then, we may well have noninvasive interventions that also help the brain to better drain plaque debris out through the meningeal lymphatic vessels, and perhaps even successful microglial pharmaceutical interventions that work without doing harm.

Very likely some of these therapies will be used in winning combinations. For instance, the researchers in Australia who recently discovered, like Tsai, that delivering light waves can clear amyloid plaques in animal models also discovered, in 2017, that when they combined a biologic immunotherapy (one which delivers antibodies to help decrease immune-activated inflammation in the brain) with therapeutic ultrasound, it significantly enhanced the effect of the immunotherapy and the reduction of amyloid plaques and tangles.

In our future hypothetical, Katie doesn't have to try all available interventions at once—she can try one or two at a time and continue to go in for regular microglial PET scans as she and her doctor monitor which therapies are working in her individual brain, given her genetic profile. Over the next year, they can put together a combined treatment plan, one that they can see is succeeding in real time—and one that will prevent the forest fire of Alzheimer's from ever becoming fully kindled in Katie's brain in the first place.

And one day, perhaps a decade from now, a teenager might be able to go in to her pediatrician's office for a yearly blood test to screen for specific signatures that reveal microglia are shifting their activity—the same way we now screen preteens for cholesterol. If the test indicates microglial or synaptic changes that exceed a healthy norm, teens and their families can consider a range of safe and early interventions.* Hopefully, by then, we'll also have even more on offer in terms of treatments and preventions for psychiatric disorders.

Remembering Alice

A few weeks later, I share this research with Katie, who has occasionally expressed concern that she might one day develop early-onset Alzheimer's, given her grandmother's history and the fact that she suffers from depression—an added Alzheimer's risk factor. Katie tells me she can't help but think of how her grandmother Alice's life might have been entirely different if she'd lived during an era when cutting-edge interventions for early Alzheimer's existed.

We are eating bagged lunches from Whole Foods at a small round green metal table in Bryant Park in New York City. Pigeons meander in the grass looking for crumbs. Katie is telling me about the ways in which her grandmother dramatically changed while Katie was growing up. She shows me a photo she keeps on her phone, taken four decades ago, in which her grandmother is a bright, photogenic, middle-aged woman. Alice stands barely five feet tall and has shiny shoulder-length dishwater-blond hair, like Katie's. She is posing proudly with her children and grandchildren on the front porch of her white ranch house. Red zinnias bloom in window boxes along the porch railings. Katie points to herself—

* As with any screening, we need to be careful, since intervening with medications that bring side effects can also do harm. In this future scenario, such testing would need to be repeated over a significant period of time, and preventative interventions would begin with approaches that are free from side effects.

she's just five years old, wearing shorts and blue sneakers. Alice stands at the center of her family, wearing a yellow summer dress, smiling, an aliveness in her eyes that reminds me of the light I see in Katie's these days. In her yellow dress, Alice beams like a small but fierce bit of sunshine.

"Nana was a piano teacher and a lot of students came to her house," Katie recalls. "She kept a glass bowl full of candy on the piano." She pulls up another photo of her grandmother, sitting with Katie's mom, Genna, and her uncle Paul on a piano bench, all three flashing grins as their three sets of hands comb the black and white piano keys in unison.

"But I don't remember her this way," Katie says, pointing to her grandmother's laughing, upturned face. By the time Katie was twelve or thirteen, her grandmother's depression, OCD, Crohn's disease, and symptoms of early Alzheimer's had started to etch away at the mind and body of the vibrant young woman we see in these photographs. "Once Alzheimer's really took hold, everything became magnified," Katie explains. "Her hoarding. Her agoraphobia—refusing to leave her house. Refusing to throw anything away. Not following up on her doctor's advice. If Nana had trouble finding something, she'd call the police, convinced that someone had broken in and stolen it. The police began to fine her for excessive false alarms." Whenever Alice's Crohn's flared, Genna would have to call 911 to get her mother to the emergency room for medical attention. "Each time the EMTs came, they had to clear a new pathway through stacks of newspapers and boxes for the stretcher," Katie says. "Nana would yell at them, cursing them for barging into her home."

In the last photo Katie pulls up on her iPhone, she is standing with trash bags outside her grandmother's ranch house. Only now, the white paint is peeling and the porch sags under the weight of piled-up detritus. An old dismantled swing set, rusted lawn chairs, bikes, and discarded furniture are stashed alongside plastic storage bins covered with tarps. Pine needles and leaves appear to have been raining down on this discarded junk for decades.

Katie and her mother routinely tried to clean out the clutter, she says, but if her mother tried to throw anything away, her grandmother would stand at the door barring her from removing it, or worse, calling her daughter "a cheat!" as if she were stealing from her. So it was easier, over time, Katie says, with regret lacing her voice, to give in and let her grandmother remain in what began to resemble "a rat's nest." Katie sighs. "My mom would go regularly and check on her, care for her, even after Nana barely recognized who she was. We had people coming every day to bathe my grandmother and bring her meals. Over the years, though, I could see how my grandmother's illness was kindling my mom's own depression and making it worse." It's not surprising, Katie says, that "while my mom was caring for Nana, she developed two autoimmune diseases of her own.

"Nana died in this house," Katie tells me, tapping the photograph. "My mom couldn't convince her to come to an Alzheimer's retirement center near us."

But what if Alice's story had been an entirely different one? Let's imagine instead that one day, during the years that Alice was still teaching piano, or smiling from her porch in her bright yellow summer dress, she had gone to the doctor. And that her doctor—in our future medicine scenario—had been able to run a quick scan that showed that Alice tested positive for very early, and worrisome, signs of synapse loss.

And what if these biomarker tests showed there was a notable synaptic loss in areas related to both depression and Alzheimer's? And what if, given these findings, Alice had then been referred to a gamma light flicker therapy practitioner for Alzheimer's, as well as transcranial magnetic stimulation for depression? (Along with, of course, other lifestyle changes in diet, exercise, and sleep.)

Would Alice still have died in her pack-rat home, refusing to leave, screaming at her daughter and granddaughter as if they were thieves stealing what was precious from her—or might the silent killer that stalked her mind have been stopped in its tracks?

A Hopeful Road Forward

It may be some time before we have such actionable interventions for Alzheimer's disease at your local doctor's office, but Li-Huei Tsai at MIT isn't waiting around—she's investigating whether gamma light flicker therapy can change the trajectory of patients' lives today. Tsai has already collaborated with fellow scientists to create a bioresearch team called Cognito Therapeutics to launch human clinical trials.

To find out more, Tsai sends me to speak with Cognito's president, Zach Malchano, who explains how his team is translating what he refers to as Tsai's "sensory stimulation technology" to human patients. "The work from Li-Huei Tsai's lab is our jumping-off point for the work we are doing here," he says. Cognito is utilizing Tsai's protocol—delivering gamma oscillations at 40 hertz for an hour a day for seven days—in a small cohort of subjects who suffer from mild cognitive impairment. "Our first task is to see if these patients tolerate gamma therapy, and our second is to see if gamma therapy will help them experience some measurable cognitive and quality of life benefits." It's a blind study, meaning patients won't know whether what they're getting is the real therapy or placebo. Researchers will follow both the group getting gamma therapy and the control group for twelve months and then compare their outcomes.

"We're starting with patients whose symptoms are mild," explains Malchano, whose team of consultants includes Alvaro Pascual-Leone, the leading expert in noninvasive brain stimulation at Beth Israel Deaconess Medical Center and Harvard Medical School. "Patients with Alzheimer's whose disease has progressed have few options at this point, so we would be excited if we were able to provide even modest benefits for individuals in the early stages of the disease that will continue to benefit them and have an effect on disease outcome later on." If Cognito's first study results appear promising, they will design a larger trial that satisfies FDA requirements for cognitive assessments for patients with full-blown Alzheimer's, and treat patients with more obvious signs of the disease with gamma light flicker therapy.

"We have to temper our excitement that we might be able to also use this not only in Alzheimer's but in other neurodegenerative, neurodevelopmental, and psychiatric diseases," Malchano confesses. "There are so many potential applications. We can't wait to see. But we are going one careful step at a time."

Pascual-Leone, meanwhile, is also beginning a new clinical trial—introducing gamma oscillations in Alzheimer patients using a technique called *transcranial alternating current stimulations*, or tACS, which, like TMS, is a noninvasive brain stimulation method. Pascual-Leone's group has already demonstrated that in non-Alzheimer's patients "we can reliably modify and increase gamma activity in humans in specific parts of the brain, which result in behavioral gains in memory."* They are now treating Alzheimer's patients to see if tACS alters microglial activation (as measured by PET scans), decreases the amount of amyloid plaques and tau protein in the brain, and confers cognitive benefits for patients.

Meanwhile, Li-Huei Tsai is already using gamma light flicker therapy on her own brain. On the day that I wrap up my visit with Tsai at Picower, she picks up the small LED light flicker apparatus she invented. She confesses that she treats herself "for an hour every day in my office. And a great many of my colleagues here at MIT are lining up outside my door to be treated too." She smiles. "This is where we are going with Alzheimer's treatment. This is the future."

---

* This research is being led by Emiliano Santarnecchi, Ph.D., assistant professor of neurology at Harvard Medical School and a clinical research scientist at the Berenson-Allen Center for Noninvasive Brain Stimulation, and is being funded by the Defense Advanced Research Projects Agency, or DARPA.

ELEVEN

Desperately Seeking Healthy Synapses

HEATHER SOMERS IS NOT SLEEPING WELL. "I'M TRYING TO RESIST the urge to nap away most afternoons," she says, filling me in on the past few tumultuous months in the Somers home. "I'm so tired. I'm finding it hard to teach. I'm no longer feeling motivated to throw myself into the same work that filled me with so much energy for years. I can't really even muster up the energy to prepare meals for my family." She pauses. "I'm sad all the time, and I am *not* a sad person."

Recently, Heather, who has gone to a partial teaching load to allow her greater flexibility, went to visit her daughter at college. "She's not doing as well as we'd hoped," Heather says. "She's going to the counseling center, taking her medications—but her panic and anxiety disorders are not getting better. She's just a baby step away from not being okay." On her way home, Heather found herself so caught up with thinking about Jane that, she says, "When it came time to switch trains halfway home, I ended up on the wrong platform. By the time I found the right track, I'd already missed my train, and the next one wasn't for two hours. I just sat on the cold metal bench on the platform feeling so depleted. It was as if I could feel my whole being sinking into this vortex of maternal worry and grief and sadness. And then I just began to sob, while other passengers walked by."

A short while later, Heather was still wiping a few tears away when she was interrupted by a woman who had a toddler with her. The woman asked Heather if she was on the right platform for the train to New York; she'd found the signs confusing and had missed her connection. After Heather reassured her she was in the right spot—and that she too had missed a train—mother and child sat down on another bench nearby, and the mother began to read a book aloud to her daughter.

Heather emerged from her funk, suddenly aware of how tenuous her mental state was. "I'm wallowing in self-pity, breaking down, and here's this woman with a baby, and she is just making the best of things. In that moment, I realized that I'm struggling more than I have been willing to admit."

Heather and I are sitting with cups of chamomile tea in a coffee shop and bakery near Jane's college campus. Heather, who grew up in Maryland, is staying with her elderly parents, who live in Baltimore. She plans to be here for a few days each week, not only to help support Jane but also to help her parents sort through and clean out their house before they move to a local retirement community.

Heather is stocking up on gluten-free nut breads, soups, and other healthful goodies for Jane's mini fridge, and for her own stay at her parents' home. "I'm trying not to do too much for Jane, but her therapist and psychiatrist also suggested she will need extra support if she is going to attend college full-time so far from home," she says. As Heather talks, the bright light coming in the café's window falls starkly on one side of her face, emphasizing the worry lines fanning out from her serious gray eyes.

Such is the dilemma that families face when their teen is suffering with a mental health disorder. There is the initial stigma a teenager confronts in acknowledging that she or he is flailing, and in asking for help in managing the frightening dark web of depression or an anxiety disorder. Then there is the secondary stigma a family feels in admitting, in a society that celebrates kids' awards, college acceptances, and team trophies, that their child is not only not doing any of those things, she is finding it hard just to get out of bed. Amid all this there is the tertiary burden that falls on a mother like Heather because of the pejorative cultural view that if a child is troubled, the mother must be in some way at least partly to blame: She must be too much of a helicopter parent, or a tiger mom, or too intrusive, too codependent, too demanding, too invalidating, or some toxic combination of the above. While bearing up under all this, a parent like Heather still has to figure out how to manage her anxiety over her child's well-being and how best to intervene and help her child without doing too much, or too little. "It's like trying to learn to dance on the head of a pin, at the very moment you feel like collapsing," Heather says.

I can see all of this weighing on Heather today, in the weary expression on her face, in the light of the windowpane. The said and the unsaid. And how this unnecessary layering on of cultural shame engenders a toxic self-questioning in her, at the very moment a mother needs to be at her most sure-footed, clear-headed, and confident in order to help her child thrive.

Really, it's no wonder Heather broke down on the train platform.

"This is an endurance game," she says. "I'm caught between wanting to give up everything to help my daughter, and my husband, and my mom and dad, and feeling angry that all of my energy is just continually sucked out of me." She sighs. "I realize I also have to find help for *me*." She pauses. "What I really think I need is a synapse doctor!"

And so Heather has decided to pursue what is, for her, a wildly novel undertaking. She's about to try a treatment that, like TMS, is slowly becoming better understood and appreciated, given our new understanding of the brain, microglia, and synaptic health.

She's made an appointment for an evaluation with a psychologist and neurofeedback specialist, Mark Trullinger, Ph.D., who utilizes qEEG neurofeedback to treat patients and who has coauthored studies on the benefits of qEEG neurofeedback with cognitive scientists at Johns Hopkins and the University of Maryland School of Medicine. "It makes sense for me, schedule-wise, to be treated here since I'll be here on a regular basis for the next several months," Heather says.

Although neurofeedback has been around longer than transcranial magnetic stimulation, it has only recently become better accepted as a form of neurotherapy by the medical community. Studies of it are also relatively new; as a field, it is in its late adolescence or early adulthood.

And that's in part because, like TMS, for years, no one entirely understood all the mechanisms by which neurofeedback can change the brain, so it has often been considered a suspect therapeutic approach. But our new understanding of how microglia generate the underpinnings of vibrant circuitry in the brain, and the scientific revelation that microglia can be potentially rebooted, are changing the way brain hacking methods are viewed. Indeed, in the past several years, as we've come to understand the brain as an immune organ—one in which microglial immune cells oversee synaptic health—neurofeedback (which helps bring underactive synapses back online) has been generating renewed scientific interest.

Researchers at Johns Hopkins have shown that individuals who've undergone neurofeedback training, combined with mindfulness training, show growth in the volume of the brain's hippocampus. Other studies have found, using fMRI scans, that those who undergo a treatment course of neurofeedback show an increase in white matter and gray matter volume in the brain in areas associated with attention. Similar studies show growth in cortical gray matter in patients with postconcussion syndrome who've undergone neurofeedback therapy.

And we know that that can occur only if microglia back off and allow neurocircuits to come back online and start to light up again.

A number of other recent studies show improvements not just in brain matter volume but in patients' symptoms. As we learned in chapter 8, in one 2018 single-blind randomized controlled trial of patients treated for depression with neurofeedback, participants experienced a 43 percent decrease in their depressive symptoms, and nearly 40 percent went into remission.* In a smaller pilot study, nearly half of patients with major depressive disorder who were treated with neurofeedback experienced remission. (The results of these studies on depression are similar to the percentage of patients who benefit from many antidepressants.) In a

---

* Researchers used a type of neurofeedback known as *real-time functional magnetic resonance imaging neurofeedback*, or rtfMRI-NF, which utilizes fMRI imaging to capture and reflect what's happening in a patient's brain during treatment so that both patient and practitioner can use that information to shape brain wave activity toward a healthy norm.

2016 randomized controlled trial, nearly three-quarters of patients with PTSD no longer met the criteria for having PTSD after twenty-four sessions of neurofeedback. Similar trials show a statistically significant improvement in symptoms of generalized anxiety disorder in patients after a course of neurofeedback treatment, as compared to patients who receive no treatment.*

Still, randomized clinical trials on the efficacy of neurofeedback need to be repeated and replicated with larger numbers of people, since most of these studies involved small groups of patients. For that reason, neurofeedback is still often thought of as most beneficial when used as an adjunct therapy, in combination with talk therapy, cognitive behavioral therapy, mindfulness training, and other interventions.

Nevertheless, these recent findings offer hope for patients like Heather. Plus, neurofeedback is frequently more accessible and affordable for patients than TMS, since many neurofeedback practitioners accept insurance, and many insurance companies cover all but a manageable copay. With the hefty medical bills that a family like Heather's is already juggling, on top of having two children in college, insurance reimbursement can make a tremendous financial difference.

Affordability matters on a psychological level too, Heather confides. "It's difficult for me to spend money on myself when we have so many bills for Jane and Dave. I don't want what I do for my health to financially limit how much we can do for them. But ultimately, I have to help myself if I'm going to be able to offer Jane the steady support she needs. I get that now."

I offer to meet her at her new practitioner's office the next day for her upcoming evaluation.

---

* Other research has demonstrated that in a significant number of patients, neurofeedback improved cognitive processing, mood, and anxiety symptoms in those with treatment-resistant obsessive-compulsive disorder; helped alleviate anxiety in patients with chronic illness; improved fatigue levels and cognition in cancer patients; and improved cognitive scores in patients with traumatic brain injuries. Randomized controlled trials have shown that neurofeedback helped to alleviate symptoms of both restricted and binge eating in patients with eating disorders, and that these positive effects remained three months after treatment had ended. Several randomized controlled trials have shown improvements in symptoms of ADHD in students.

Reading Heather's Brain Waves

Heather and I are sitting in the white-walled, sparsely decorated waiting room of Dr. Mark Trullinger in Lutherville, Maryland. Trullinger, who holds his Ph.D. in psychology with a focus on neurofeedback device interventions for cognitive and mental health, founded his clinic, Neuro-Thrive, with his partner and wife, Deepti Pradhan, Ph.D., to treat patients with concussion, cognitive and attention issues, and psychiatric disorders. The wooden coffee table in his waiting area is scattered with magazines—*Neurology Now*, *Harper's Bazaar*, *American Psychologist*, *Baltimore's Child*—and large-print editions of *Crosswords* magazine. Trullinger sees patients from every age group and walk of life.

Heather signs in at the front desk. After she puts down the pen, she shakes out her right hand a few times. "I'm not sure if I'm just so fatigued, or if it's my rheumatoid arthritis acting up, or both, but my hands are so sore and tender," she says, massaging her fingers. "They have been for months."

A few minutes later, Heather is seated in a patient room. Trullinger, who has a round, boyish face and stubby blond hair and wears wire-rimmed glasses, sits facing her. As he asks Heather about her most pressing concerns and symptoms, she unspools the most salient details of her unrelenting stressors at home, as well as her health history. She wraps up by confessing, "All my ruminating about my life affects my quality of life. I use all the coping tools I've worked so hard to develop, and they help, but then my feelings start coming up out of this hole that I keep trying to stuff them in. When they come up they feel unmanageable. I'm starting to think that maybe I have post-traumatic stress from helping my family through their issues."

Trullinger nods as Heather talks, interrupts occasionally to iron out details, and makes clinical notes from time to time. He explains to Heather that in order to run a functional brain map, he'll place sensors—electrodes—on her scalp, which will interface with the computers and monitors on the small portable desk at which he sits.

In a similar process to that Dr. Hasan Asif used when evaluating Katie Harrison, Trullinger fits a yellow cap snugly over Heather's head, at-

taches nineteen electrodes, checks to make sure the attachments are secure, then runs a qEEG scan of Heather's brain waves while asking her
to follow various prompts.

In order to track and record brain wave patterns, Trullinger also utilizes the sLORETA program, or standardized low-resolution electromagnetic tomography, which produces a 3-D "brain map"—a dynamic brain
scan that is compared to a healthy brain map.

This dynamic brain scan shows where electrical brain wave patterns
in specific areas of the brain differ from those in healthy individuals.
These deviations have been correlated, through studies utilizing fMRI
images, to psychiatric and cognitive symptoms, in the same way that disruptions in proper gamma wave activity in the brain's hippocampus have
been correlated to memory problems and Alzheimer's disease. For instance, patients who experience depression, difficulty in regulating or
processing emotions, and low motivation display abnormal alpha wave
activity in the left prefrontal region of the brain, and a lack of neural connectivity between the left side of the amygdala and regions of the prefrontal cortex. Patients with generalized anxiety disorder, like Heather,
show abnormalities in the alpha and theta waves in the occipital area of
the brain, and so on.

Treatment in qEEG neurofeedback differs greatly from treatment in
TMS. To help change brain wave activity, the qEEG neurofeedback
practitioner uses different forms of brain training methods to coach and
reward the brain as it produces healthier brain wave activity. Some neurofeedback practitioners reward the patient by using more pleasing visual
imagery as the brain changes in positive ways; others cue the brain toward
healthier activity by rewarding it with pleasurable sounds.

In TMS, however, the practitioner delivers magnetic pulses through
the skull, into the brain, to help change brain—and microglial—activity.
Often, the two practices are used in tandem.

Trullinger continues to collect data on Heather's brain waves and
neural activity in varied areas of her brain—evaluating her brain waves
with her eyes open and closed—for nearly an hour. After he unhooks her
from the equipment, he asks Heather to spend some time online that
evening and fill out a series of lengthy questionnaires, which will shed

more light on how she is managing in daily life in terms of her mind state, cognitive functioning, and mood. At her next appointment, he will have put all of this data together—their in-depth conversations about her struggles, her brain scan, and data from her online questionnaires—in order to create a comprehensive treatment plan.

Later that week, Trullinger explains to Heather that her qEEG scans really lit up in her brain's parietal lobe, showing an excess of alpha waves, which, he explains, "we might associate with a state of chronic rumination, as well as with difficulty in accurately perceiving and interpreting what's going on in the world around us." This makes it hard to respond well under stress. Heather's brain map also shows specific wave patterns that likely emanate from the amygdala, which indicate that Heather is often functioning "from fight-or-flight mode."

None of this is to imply, Trullinger adds, that the stress Heather faces isn't 100 percent real. "You are reporting an appropriate amount of concern for your situation, which is highly stressful," he tells her. But her current brain circuitry and brain wave patterns indicate that there is more agitation and rumination over these worries than is healthy in day-to-day life. These brain wave patterns, he says, "can also lead us to interpret bodily functions differently and intensify our awareness of pain." This, he says, may play a role in exacerbating the debilitating stomach pain and joint pain Heather describes. Again, it doesn't mean that this pain, and her experience of it, aren't real. Rheumatoid arthritis is a painful disorder. It simply means that her perception of pain may be amplified.

In this sense, he says, Heather's brain's overactive state of rumination and agitation is "a global issue that not only affects mood, but is also present with your experience of pain and physical autoimmune symptoms."

Trullinger also points out that Heather's delta waves are very slow. There is also a lack of theta waves in "the deeper brain structure." This tells Trullinger that activity in the hippocampus is more quiescent than it should be.*

---

* This finding is based on the algorithm, or brain map, created by the qEEG computer program sLORETA rather than by measuring surface brain waves.

When theta waves are this slow, Trullinger says, "there is also a greater likelihood of processing and memory issues." This finding surprised him, he says, because although Heather reported some issues with focus due to her anxiety, she did not report any cognitive processing issues during her intake. But her functional brain scans clearly tell Trullinger that "we are looking at some cognitive processing, attention, and working memory issues that are not insignificant."

"Ideally, we would like to see that change," he tells her. However, he adds reassuringly, he can use neurofeedback "to help you process at a speed that will allow you to move in and out of working memory while retaining greater focus, and raise your metacognition so that you can shift your lens more quickly to perceive and interpret your world with less intensity." He explains, "Helping your brain change the process by which it processes and responds to things will help you then approach the problems and stressors you face quite differently." In other words, life may not be easier, but life will *feel* easier. He also believes she will experience some degree of relief from pain.

Heather rakes a hand through her short dark hair. "Can you do all that?" she asks. "I mean, I've been doing so much therapy, yoga, mindfulness, and my brain is still *this* much of a mess—"

"We don't use neurofeedback in place of therapy and mind-body practices," Trullinger says. "By training down your brain's alpha waves and increasing theta waves, we hope to help decrease rumination while increasing attention, processing ability, and working memory. This prepares the brain better for modalities like talk therapy and mindfulness so that these modalities can have their full benefit." The patients who have the most success with cognitive behavioral therapy (CBT), dialectical behavioral therapy (DBT), and talk therapy, he believes, "do it in combination with neurofeedback."

Trullinger gives a small smile. "The brain really *prefers* being better self-organized and functioning better. We just have to help it a bit in order to engage its long-term potentiation."

He thinks that in Heather's case, treatment will take twenty-four to thirty-six sessions.

Heather is game.

Hacking into Microglial Connections

So, exactly how does neurofeedback help to calm down overactive microglia and benefit patients who are suffering? I turn to one of the foremost researchers and experts in qEEG neurofeedback and training, Jay Gunkelman. Gunkelman, past president of the International Society for Neurofeedback and Research, was the first EEG technologist to be certified—in 1996—in the quantitative analysis of EEG, and he has evaluated more than half a million brain maps. He helps me to understand the link between neurofeedback and microglia when we talk by phone from his home in California. Gunkelman connects the dots this way: All brain waves—beta, theta, gamma, alpha—are regulated by a slower pulse, and a lower frequency, that's constantly produced and emitted by the underlying electromagnetic field in the brain, known as the direct current or DC field.

Outside the body, DC fields are created when you rub two objects together in such a way that electrons build up on one of the surfaces. In everyday life, you can see this when you rub two blankets together in the dark and sparks fly, or if you rub a balloon against your hair and your hair stands on end. We call it static electricity. DC fields also exist in the atmosphere around us, such as during a thunderstorm.

In the brain, says Gunkelman, the activity of this direct current field is regulated by the activity of glial cells.

The direct current field loosely represents the brain's state of health. Individual brain waves cannot be performing ideally if the direct current—the underlying slow-pulse electrical frequency that regulates the brain—is not functioning as it should. And the direct current cannot be in good form if microglia are not behaving in a balanced and measured way.

When microglia are overactive, as we've seen, they create disruptions in synapses and a loss of neurocircuitry. This loss of circuitry can be measured by the brain wave patterns that emanate from the underlying DC field. "Using a noninvasive charge such as neurofeedback, which helps regulate the DC field, allows us to influence the overactivity of microglia," explains Gunkelman. And this in turn can help to improve a patient's mood and cognition.

Among other happy by-products, one treatment benefit of neurofeedback, especially for patients like Heather, is a decrease in sensitivity to pain.

Microglia and the Pain-Perception Feedback Loop

Pain is a complex neural-microglial-emotional experience. When an area is inflamed—red, hot, painful, swollen—as in Heather's case, in rheumatoid arthritis, pain signals can be protective, making sure an individual doesn't overuse affected joints or muscles. But when pain persists over time, in the case of physical injury, overuse, and also in some autoimmune disorders, it can begin to restructure neural pathways in the brain in ways that can worsen pain sensations. This happens—once again—because of microglia.

In 2015, researchers at the University of California, Irvine, and in Canada showed that nerve injuries and chronic pain sensations activate receptors on microglia that can, in turn, spit out neuroactive substances that block the normal "reward" signals that tell us our body is healthy and well, while also fueling neural pain networks. This can lead to the creation of new neural pathways that shoot us more pain signals, slowly changing the nervous system over time to make it more sensitive to future painful experiences. More pain in the body sets off more pain pathways in the brain, creating a "kindling" effect that maintains and worsens pain sensations.

Also in 2015, researchers at Massachusetts General Hospital at Harvard Medical School compared PET scans of the brains of those with chronic low back pain with those of healthy volunteers and looked for signs of activated microglia. Patients with chronic pain had much higher levels of proteins associated with activated microglia; in fact, researchers could tell simply by looking at an individual's brain scan whether he or she was a pain patient or a member of their healthy control group.

Neuroscientists have also found that when "accelerated" microglia kindle pain pathways in the brain, this in turn restricts the release of dopamine, a neurotransmitter that helps control the brain's reward and pleasure centers. These neurotransmitters help to regulate not only pain but also our mood. As microglia congregate in and dismantle the brain's pain-moderating capabilities, it creates a cascade that changes the reward center in the brain, making it more difficult to experience pleasure. The fact that microglia actively contribute to the establishment and maintenance of persistent pain by congregating and causing inflammation in brain centers that moderate pain, pleasure, and mood helps to explain why so many patients like Heather, who experience chronic pain disorders, also face much higher rates of depression and anxiety—and vice versa. (It may also explain why long-term sufferers of chronic pain syndromes have such frighteningly high suicide rates.)*

In 2015, Japanese researchers discovered that in patients with rheumatological autoimmune disorders, like Heather (as well as Katie's mom, Genna), systemic inflammation stemming from a bone disorder signals microglia in the brain. This in turn sparks neuroinflammation, which contributes to these same patients' greater likelihood of developing Alzheimer's later in life (much like Heather's grandfather and Katie's grandmother Alice). Once bone inflammation stokes brain inflammation, overactivated microglia signal inflammatory chemicals that intensify patients' sensations of physical pain in the body.

---

* One more interesting piece of research underscores how pain and emotional mood states are interrelated: In one study, researchers gave half of study participants acetaminophen (the active ingredient in Tylenol, most often used for physical pain relief, such as toothache, headache, or menstrual cramps) and the other half a placebo. Participants took the pills for three weeks without knowing whether they were receiving the drug or the placebo. Then researchers had all the participants take part in a virtual ball-tossing game as well as fill out surveys that measured their degree of emotional pain, responding to statements like "Today, being teased hurt my feelings." Those who'd been taking acetaminophen felt less emotional pain when they were left out of the ball-tossing game or teased. Treating pain with acetaminophen seemed to reduce not only pain, but also the sting of rejection that people felt at being left out. (That's not to encourage patients to pop acetaminophen—not at all. Acetaminophen has been linked to GI problems and, in large doses, liver failure. Additionally, other studies have shown that acetaminophen also dampens people's sense of empathy, which is usually not a good thing.) The point here is this: We are still a long way from fully grasping how intricately pain pathways and emotional pathways in the brain are intertwined.

. . .

So, given all that, I am curious whether neurofeedback can be potentially transformative for a patient like Heather, whose daily life is—and
will continue to be—full of high-octane stressors, given her complex
multi-caregiving roles, her own serious illness, and her chaotic dual-city
life. After all, as Trullinger has noted, Heather's distress is, in many ways,
to be expected given her situation. Can neurofeedback help create
changes in her brain that will persist long into the future, even as these
stressors continue to tax her in body, mind, and spirit?

I place one more call to another leading expert on neurofeedback,
Sebern Fisher, M.A., founder of the Optimal Brain Institute in Northamp-
ton, Massachusetts, and author of one of the industry's bibles, *Neurofeedback in the Treatment of Developmental Trauma: Calming the Fear-Driven
Brain*. When we speak, Fisher sums up the power of neurofeedback to
create long-lasting changes in the brain this way: "The amygdala, which
is devoted to our survival, is not all that intelligent," she explains. "It is
just reactive to stimuli that it considers dangerous. It can be hard to up-
date it," especially in those who've experienced a great deal of stress and
trauma. In these patients the amygdala is often in a constant state of
threat assessment, trying to detect the next danger coming in the next
moment, and the next. "There is also a hyperlinkage between the amygdala and an area of the brain known as the periaqueductal gray, or PAG,"
she explains. The PAG sits near the brain stem, and in patients who've
faced chronic stress, "it just keeps stimulating the amygdala over and
over again. The nervous system cannot quiet itself—and that changes
the behavior of microglia, which changes the brain's immune system."

In patients who've also experienced trauma early in life, or adverse
childhood experiences, pathways between major hubs of the brain that
need to communicate also appear to be abnormal. They aren't as stable—
which makes communicating between crucial areas of the brain much
more difficult.

Neurofeedback, Fisher contends, helps the "amygdala to recognize
it's safe now, so the brain can learn to be calm again." And in patients
who grew up with adversity and trauma, it also helps pathways between

major hubs of the brain to communicate with each other more efficiently again.

In fact, adds Fisher, in her recent collaborations with some of the most famous meditation leaders in the world, they have come to believe that neurofeedback can serve as what they now refer to as a "new Dharma door."

As mainstream medicine begins to increasingly understand that having a healthy brain is "more about circuits and less about chemistry," she says, neurofeedback is becoming more mainstream. "Changing how these circuits fire—their bad habits as it were—is what neurofeedback does," she adds. "The brain is devoted to its own optimal function. It has to be. These symptoms in the mind or body are indications of miscuing in the brain. Neurofeedback isn't a miracle—it's just better than anything else I know of."

I share all of this with Heather, who confesses that she doesn't fully understand the science of how and why neurofeedback works, but to her it's worth undergoing the twenty-four to thirty-six sessions Mark Trullinger suggests for her treatment, to see if it can make a difference.

"We still don't know why or how some medications work, or why in some people medications cause such terrible side effects," Heather muses. "So the fact that I can't fully grasp the mechanism is okay. At least this has zero side effects! I'm in."

TWELVE

Rebooting the Family Fixer

H OW IS IT THAT A BALLERINA CAN EXECUTE MULTIPLE PIROUETTES without experiencing vertigo and falling down? How is it that a concert pianist can play Bach or Beethoven flawlessly—even with his or her eyes closed? Neither the ballerina nor the pianist was born with these skills. Neurocircuitry in their brains has slowly changed over time to allow them to do what their brains could not manage to do before. Their brains have been trained to work very differently.

Neurofeedback retrains, self-organizes, and upgrades the brain in new ways too.

Heather's First Treatment

Mark Trullinger's instructions to Heather, before she undergoes her first neurofeedback treatment, are very simple. "Every time you hear a beep, simply tell your brain, 'Good job!'" Each time Heather hears a beep, he explains, it means that her brain waves are approaching a more ideal state—lower alpha, higher theta—which are correlated with improved cognition and focus.

"The more beeps you are hearing, the more we know that your brain is learning what we want it to learn. Each time you tell your brain 'Good job,' it helps to reinforce your brain's proper functioning. Don't try to force anything. Just lie back and try to stay in a state of mind similar to how you might be if you were to fall asleep in front of the TV, but you still hear the TV in the background. Just listen for the beeps, tell your brain 'Good job,' and breathe, don't forget to breathe."

Other than that, Heather need do nothing at all.

Heather lies back in a chair as Trullinger places six electrodes on her scalp and connects them to his computers. In TMS, practitioners often run a live dynamic brain scan during treatment because they need to see where the brain is overactive or underactive before delivering a magnetic pulse. In qEEG neurofeedback treatment sessions, practitioners usually hook up electrodes only to the specific area of the brain they are treating, and they may need only one to six electrodes to successfully monitor that area during a particular session.

After Trullinger sits down in front of his computer to monitor Heather's treatment, we begin to hear a brief, low beeping sound every five or ten seconds. To my ear, the beeps sound like a tiny dog going *Woof!* . . . *Woof!* And so, I think to myself, this is not unlike training your dog to bark at unwanted intruders (or, in this case, unwanted brain wave patterns) and saying "Good job" each time they do.

Only in this case, Trullinger is helping Heather train theta waves (which signal activity having to do with focus) to jump higher, and alpha waves (which are related to more of a daydreaming state) to arc lower. When Trullinger, who sits at his computer screen, sees these brain waves move closer to each other, he rewards them with that beep and Heather rewards them with a thought. He continually adjusts the tones—or rewards—to come more frequently as the theta and alpha waves get closer to each other, slowly shaping the brain. "I teach the brain waves to keep making that pattern by using this specific beeping sound," he says.

Trullinger explains that the brain registers the sound as a reward because Heather has been told that the beep is good, and she tells herself it is good. In her mind she desires to get more beeps because she believes it will improve her brain waves. This belief will get reinforced when Trullinger shows her changes in her brain waves at the end of each session and again as Heather's symptoms start to change.

It's a little like paper money, Trullinger says. "If we did not have context for the value of a hundred-dollar bill, then it wouldn't mean anything, just like the beep doesn't. But once we know that hundred-dollar bills get us more of what we want in life, allow us to survive, and are generally thought of as a good thing, they become a reward. When the context is controlled around something and it is established, like a beep, it can become conditioned. This is not entirely operant conditioning—it uses social learning, associative learning, and to some extent aspects of classical conditioning to establish the beep as a reward."

Imagine, for instance, that you're lost while hiking in the woods—and you're trying to find the right trail. If you repeatedly keep taking the wrong pathway, one that leads you in the wrong direction, over time that trail gets more trodden, better marked, and more clearly defined. After a while, you can't see any other way through the terrain, even if you know this one won't serve you well. But once your brain learns to recognize another and better trail—one that is far more efficient and helpful to you, given where you hope to go—it starts to be easier to find that new pathway now and in the future. The more you take the right pathway, the more defined it becomes. In that sense, neurofeedback can help give your brain a new map for how to function more optimally in the world.

Each time Heather's brain takes the better path, Trullinger delivers a beep. After half an hour, the beeps start coming closer together. And then that's it. Heather's first treatment is finished.

"We can already see that your alpha and theta waves have crossed several times and they are moving closer to where we'd like them to be," Trullinger tells Heather. "What we are looking for here is to reward the brain when alpha and theta waves begin to cross over each other in just the right way." He shows us an image that looks like two sets of valleys and hills, superimposed on each other, only as one hill is rising, an undulating hill from the second set is descending, crossing over the first. Together, the crisscrossing waveforms form an awkward, almost caterpillar-like shape.

Heather looks at him. "And that means?"

"It means we are already making good progress," he says. "Which bodes well for treatment."

Treatment Six

After Heather's sixth session with Trullinger, we grab another cup of tea. Heather confides that she is still caught up in worrying about everyone else, but, she tells me, she might be noticing a few subtle changes.

"I know that something small is shifting because I feel a little different when I am with Jane," she explains. "That's my greatest trigger. And if I'm not as triggered when I'm with her, that is a telling barometer for me."

She recounts events of the previous week. Jane had just arrived home for fall break with a sprained ankle and it wasn't healing well. She was in a lot of pain and on crutches. Heather had made her an appointment with a specialist. "Before she came home, Jane had asked me to relay a few questions to the orthopedist we were seeing, and I agreed, since she couldn't call during the day when she was taking midterms. I thought I'd gotten everything answered, but she was upset with me that I hadn't gotten more information." Heather sighs, pressing her thumb and forefinger to her temples as she smooths her hair back from her face, as if delivering a brief acupressure massage across her scalp.

"Maybe I was just too tired to recall all the information I'd gotten," she says. "I don't know. I resisted the urge to call her on her sassiness and just tried to see where it was coming from: her worry. I validated her feelings of fear and frustration and told her, 'We'll have to see. We'll know more after we see the orthopedist.'"

Then, says Heather, "I just sat with her in silence because I had nothing more to say that was productive, helpful, or reassuring.

"I have waited out silence from many groups of students," Heather goes on. "It never bothers me. But I've never been able to wait out silences with Jane. I've always tried to fill the silence with something helpful. Fix it. Make everything better with words. And in that moment, as I sat there with her quietly, I realized, I am *calmer*. I'm just not as agitated by the things that are usually guaranteed to agitate me."

Jane curtly asked her mom why she wasn't saying anything. Heather told her, "I don't have much more to add. I'm here for you, and we will get through your appointment together." They were just lying in "Jane's uber-comfy bed, as we often do, side by side, talking through things," Heather says.

Jane was furious. She told her mom, "Then why don't you just go! There's no point in your staying in my room!"

Heather kissed Jane and left. "Normally after this kind of altercation with her I'd be so upset. But I was so . . . oddly calm, knowing I had done all that I could do."

Heather walked the dog, then went to the grocery store. "As I walked up and down the grocery store aisles I thought of how, usually, after Jane erupts, I carry the pain of those moments around with me for hours and hours, wondering what I might have said or done differently. But I realized I wasn't doing that."

She leans back in her chair. "I am feeling calmer. Six visits has changed something in how I respond to things that normally trigger me. I see Jane's anxiety and anger and mood changes with a greater sense of detachment. Maybe I'm stepping back with a little more perspective."

Treatment Nine

"The machine doesn't lie, I guess," Heather tells me over the phone. "Or I guess I should say you can't fool the machine."

Heather is describing a recent session. She was listening to the beeps during her treatment, coordinating her deep belly breathing with her "silent 'Good job!'" But she couldn't get into her normal relaxed state. Several things were bothering her.

During this year, in which she's been working on a partial teaching schedule so that she can help her daughter and parents, Heather was also hoping to find time to lay the groundwork for a new wellness curriculum for students. She was specifically interested in developing a program to help her school better address the negative influence that social media has on teenagers' lives. It's something she's long been passionate about, and she hopes to bring it into multiple schools once Jane and her parents are in a better place. She's had time on her train rides to start sketching out plans for the wellness course, but she just hasn't, she tells me, "had the self-discipline to set a schedule for making calls and getting some basic steps accomplished.

"I'm starting to realize that I have no attention span!" she continues. "Now that neurofeedback is helping me with my anxiety, it's like a layer has been peeled off and instead of feeling anxious and crappy all the time, now I can see how unfocused I am." Heather confesses that when Trullinger first diagnosed her with having attention issues based on her interview, reports, and brain scans, she didn't really believe it.

Now, she says, "He was right. I'm able to observe how distracted and impulsive I am, in ways that keep me from focusing. I'm always pulled in a thousand directions. I can see that my distractedness is a coping mechanism to keep me from thinking about stressful things and it's not serving me." She sighs over the phone. "Even though I'm less reactive, I'm still finding it almost impossible to identify priorities, stick to them, manage my time well, not take care of everyone else at my own expense, and find a way to do the things I want to do for myself."

During this last session she couldn't stop her ruminating thoughts. "I kept thinking, *Why aren't you sketching out your new wellness course, Heather? Isn't that also one of the reasons you started teaching on a partial schedule? Why can't you see your ideas through to fruition? Yes, having several family members with health issues is hard. Yes, people need you, but you are trying to focus on what you need—so why can't you?*"

And then, she says, "I think I just kind of dozed off."

After her session, she tells me, Trullinger showed her a chart, which affirmed for her that this session was different. There was no caterpillar shape of crisscrossing brain waves.

I remind Heather that healing is not a straight road. It's more like crossing the Swiss Alps—you go up and down and up and down while generally, over time, moving in the right direction. And she is most certainly doing that.

Treatment Fourteen

Heather and I catch up on a Thursday afternoon by phone after her fourteenth session. She tells me that recently she and Jane went to a yoga class together in Baltimore. Afterward, Jane was hungry.

"There was a lot of traffic and all the food places were closed, and Jane was getting really upset," Heather tells me. "I said, well, we'll double-park and you run in and get what you want."

Jane didn't want to do that; she was feeling too anxious to go into the store alone while leaving her mother double-parked on the street in a no-parking zone.

So Jane did what she often did when she felt overwhelmed and panicky. She yelled at her mom.

"I just didn't feel that usual stab of anxiousness and discomfort in my body and mind as she was off-gassing on me," Heather says. "I used to get this gut reaction and I'd have to run to the bathroom whenever someone in my family was not okay. But instead, I took a breath and said to myself, I'm not going to engage, because at some point this will all blow over. In the end, this will work out. This is just what is happening right now. It just *is*."

"Your stomach isn't bothering you anymore?" I ask, reminding Heather that on the day we first met, she described her stomach pain, GI problems, and having to race to the bathroom as some of her most debilitating symptoms. They were interfering with her ability to live a normal life.

"It's not bothering me anymore," she says. "And now that it's not, I'd have to say, wow, looking back, that really affected *every* part of my day, and yet I really didn't talk about it with anyone. I think a lot of people with anxiety have stomach issues and don't talk about it." She goes on, "There are not many treatments where the benefit is that you find yourself going about your day with better emotional energy and get physical benefits too. So that you can start to put your energy into the right things again." Heather gives a small laugh. "Even with talk therapy, I can walk out and I just feel exhausted and overwhelmed. But with neurofeedback, there are no negative effects after a session. Just tiny benefits that are beginning to build over time."

Treatment Eighteen

The real test of how well treatment has been helping Heather came when she received a call from the dean of medical affairs at Jane's college letting her know that Jane had just had several severe panic events at school. Heather talked to Jane, who was feeling better by then, but mother and daughter agreed it would be good if Heather could come down.

Heather was in Connecticut and decided to drive to Baltimore. She arrived in the late afternoon. "I just walked into the room and hugged her so hard," Heather recalls. "Something in my mind was very clear in a way it wasn't before. I was very aware that my job was simply to be there to help her feel safe. That was it." In the past, Heather says, "I'd be reacting with panic, so wrapped up with what-ifs, to the extent that I wouldn't be able to see what she needs, what she is feeling. Or I'd be worrying about what I was saying, wondering if I was saying the wrong thing to her, feeling bad about not saying things perfectly. But I was just very clear, that I needed to help Jane feel safe, and hold her."

Jane cried in her mother's arms for the first time in years.

Heather pauses. "It has only been nine or ten weeks since I started treatment. But I'm in a place I've never been in before, where even as stressful things are happening, I've somehow managed to step outside of the messy middle." And that makes her a better listener, Heather says. "Because I'm not as reactive, I'm doing a better job of listening to other people too." She is not the only mom in her cohort of friends, she says, who is having major issues with her teenage or young adult child. "I'm finding that even though it's not something people like to talk about, a few of my friends are becoming really open with me about their struggles in a way they didn't used to be, and that's because I'm better able to just be there for them without trying to fix it, or jumping in to share my own situation. It feels good, to be able to be that person who can really hear and reflect on what other people are saying and not be thinking about myself." She pauses. "I like this new me."

Later, Heather noticed something else that was new. "I'm starting to feel that my focus is better. I'm not struggling with small decisions. Usually planning an emergency trip like going to see Jane at the last second would mean I'd be very scattered, packing my bag, forgetting things, even forgetting to get train tickets. And then I'd be beating up on myself for being so disorganized and forgetful. But making all those last-minute arrangements felt . . . easy. Not as overwhelming." She pauses and adds, "The daily mental grind feels less . . . grinding."

Part of this, she says, is being better able to determine, and tell other people, exactly what she needs. For instance, after she got the SOS call from Jane, "instead of pretending and protecting Dave and handling everything by myself and keeping my fears to myself, I told Dave, 'Look, here is what's happening. I'm feeling very concerned and I'm going down. I need you to take care of hiring someone to feed the animals, and pay this tuition bill due for Ian, which I'm leaving on the counter, and I'll call you when I get to Baltimore.'"

In the past, Heather adds, in a reflective tone, "I guess I would have let people know I needed help in a passive-aggressive way." She pauses and gives a small laugh. "I would have had to go hide in the tree house as a way of showing everyone, 'Hey, I'm not okay.' But now, I can just say, 'Hey, I need your help.'"

Treatment Twenty-Two

"My hands aren't aching all the time," Heather tells me when we talk next. She's noticed that her overall sensation of pain has been improving, incrementally, for the past several weeks. "I'm realizing how many things I've avoided for years—writing by hand, typing, holding the dog's leash, cooking—because of the pain in my hands. But I'm getting back to doing all those things."

Heather and Dave recently adopted a second dog. "For the longest time, I just didn't think I could handle a puppy, physically, even though I wanted one. But I realized that I *can* handle it now. I'm in less physical pain. I don't think I would have even considered getting a puppy without having done the neurofeedback."

At the same time, she's also learning to let go of some of the things she no longer wants to feel she's single-handedly responsible for. "I've come to accept that I can't fix everything for everyone," she explains. Heather recently told her parents that they needed to hire help to sort through and box up all the "stuff" in their house — she can't do it all, and moreover, she doesn't want to. So her mother and father have hired a team of packers and Heather is organizing them as they all work together.

"I can't always make magic happen," she adds, letting out a long breath. "Especially Jane. I can be present, and ride the waves with her, but I can't save her from her panicked thoughts. She's going to have to get there on her own, with our support of course — but it is *her* journey."

Heather recently suggested neurofeedback to Jane. Jane has agreed to try it during her spring term. "She sees how much better I feel when I'm self-regulated, and how much neurofeedback has helped me," Heather says. "I can see that by helping myself, I've actually helped Jane see that she has more treatment options."

The Final Treatment

Heather has taken down the tree house. "Yes, *that* tree house," she says, cradling a tea mug between her hands to warm them as she sits in my kitchen. "It's been falling apart, and nobody has been in it for years, except for my little visit last August." A few weeks ago she asked a handyman who sometimes does yard work for her to remove it. The next day, down it came.

"I had planned some ceremonial goodbye," Heather says. "But I was unusually busy that day and had a meeting with administrators at my school to talk about my ideas for our wellness curriculum on social media, with another teacher I'm teaming up with." So when the tree house came down, she says, "Nobody was home. Nobody witnessed it. It held so many memories, and yet it was just unceremoniously removed along with the fallen sticks in the yard."

Heather says that now, when she looks out the kitchen window, "It's gone, and, honestly, the yard looks much better. Maybe it even helps me to see, visually, how time has moved past the twins' childhood. The days of tree houses, swing sets, tents, forts, dance recitals, bikes, scooters . . . they are over. But so too are the heavy-lifting years of parenting."

For a long time, Heather explains, "I did not seem to be able to let go of their childhood, emotionally. To look in the mirror and see who I am, what I want. And now I'm ready to do both of those things."

Meanwhile, Heather is excited about her proposed social media and wellness program for adolescent girls. "I'm working with administrators who really want to help anxious girls, and we are developing a really exciting program." I can hear the excitement in her voice. "I'll be running a class to train teachers. I'm getting so much work done. I feel like I have a new set of more focused brain responses."

She's made a few other changes too. "I sold my big SUV gas guzzler and bought a little used Prius," she says. "I've wanted to do that for the longest time."

I ask her how she would describe herself now, versus before she started seeing Mark Trullinger for qEEG and neurofeedback.

"I'm not apologizing all the time, because I have more self-compassion," she says. "And I'm more aligned with my own goals and purposes, and how to accomplish them. I'm taking care of myself. I'm in bed by ten o'clock, I'm meditating every day. I take time every day to walk and play fetch with the dogs. I whiz around in my little car. I just feel so much joy doing these things. I didn't grow up feeling I was entitled to joy. And now I feel I am. This is a new level of self-care.

"My new motto in life is that if it's not happening right in front of me—or to the left or the right of me—I am safe."

Sometimes, Heather says, "I think of calling Mark Trullinger and asking him, 'What did you do, exactly? I'm not the same person.'"

. . .

Later, I share Heather's thoughts with Trullinger. He explains the changes that Heather is experiencing in both neuroscientific and almost poetic terms. "Over time, Heather's alpha and theta waves started to naturally alternate in a gentle flowing manner, as they should," he says. "Her amygdala function quieted down from a shout to a whisper. And her hippocampus increased its function, becoming a more powerful, resonating force rather than meekly whispering."

Trullinger underscores, "We did not change Heather's personality, nor did the neurofeedback." Instead, he sees it this way: "When the brain is not functioning correctly, it robs the person, everyone they interact with, and the world of their true personality. It covers it up. So the goal is always to help sculpt the person's brain to help them overcome an inadequately functioning brain."

Heather says she recently looked out at where the tree house used to be in the yard and drank in the sight of the sunlight as it dappled the green, flickering leaves of the trees, which were once blocked by the old, beleaguered wooden fort. "I'm really so far from that morning hiding in the tree house," she says. "I'm building something entirely new."

THIRTEEN

In Search of a Fire Extinguisher for the Brain

I N 1991, ALAN FADEN, THEN CHIEF OF NEUROLOGY AT THE VETERANS Administration Medical Center in San Francisco, made a monumental career decision. He'd been both a clinician and a researcher for sixteen years and had helped hundreds of military vets who were suffering from traumatic head and spinal cord injuries during his tenure at several top-tier military medical institutions. For years, early in his career, he'd worked part time in community hospital emergency departments, relishing the challenge of emergency medicine while also lending his military expertise to helping to treat kids and people who'd suffered debilitating traumas.*

Still, it was a frustrating field to be in at the time: After seeing patients for the initial injury, physicians had very little to offer them in terms of how to avoid long-term repercussions from spinal cord and head injuries.

So after sixteen years as a practicing neurologist, Faden decided to shift his focus from primarily clinical work to exclusively pursuing research. First, he took a position as dean of research at Georgetown Uni-

---

* Faden spent his early years as a neurologist at Walter Reed Army Institute, focusing on shock trauma and spinal cord injury, before becoming vice chair of neurology and director of neurobiology research at the military medical school (USUHS), where he was a founding faculty member.

versity; then, eighteen years later, in 2009, Faden became the director of the newly created STAR, the Shock, Trauma and Anesthesiology Research Center at the University of Maryland, the first research center in the United States dedicated exclusively to the study of trauma, its complications, and how to prevent them.

Faden's reason for moving from being a clinician-researcher to focusing solely on research was simple, he explains, as we sit on black quilted leather chairs in the private conference room in his office suite. His office is situated not far from his several extensive lab spaces, which together total eight thousand square feet, span two buildings, and bustle with thirty-five researchers, postdocs, and techs—all working on finding answers to the concussion problem.

"Most research in mild traumatic brain injury, including concussion, has not been focused on the underlying mechanisms in the brain that lead to long-term negative consequences for patients," says Faden, who has short gray hair that runs white along his temples. "After forty years as a clinician, it was clear to me that there was still so much that physicians didn't know about how to help patients *after* they left the hospital." For instance, he says, "Maryland Shock Trauma, here in Baltimore, is recognized for having the best patient outcomes of any trauma center in the world. But we don't understand why some patients develop long-term disabilities such as cognitive decline and mood changes." It seemed to Faden that the cellular pathways involved in neurological dysfunction after head injury were still poorly understood in modern medicine, and the field needed to place more focus on research in order to address the long-standing medical conundrum of why even mild head injuries can be so life-altering in so many patients.

Faden's efforts have more than paid off. Over the past three decades, he has been a leader in the field of traumatic brain injury (TBI) research, and his group has made new discoveries that have far-reaching consequences for us all.

When most of us think of TBI, we probably think first of football players and military vets. Perhaps we've seen headlines reporting that autopsies performed on NFL players' brains show that as many as 99 percent of them have suffered from a degenerative brain disease caused by repeated blows to the head—a syndrome known as chronic traumatic encephalopathy, or CTE—which leads to memory loss, depression, confusion, and dementia, even many years down the road.* Military vets—like Heather's husband, Dave, who experienced a concussion after an IED went off, throwing him from a jeep—are also at heightened risk, of course; more than 20 percent of military personnel deployed to war zones sustain traumatic head injuries.

Indeed, according to modern biographers of the writer Ernest Hemingway, repeated concussive insults are thought to account for the memory loss, rages, headaches, and paranoia that grew progressively worse over the course of Hemingway's life, decades after he suffered concussions during World War I as an ambulance driver and later as a reporter in World War II during the London Blitz. (Hemingway also liked to box.)

But the media focus on CTE in famous athletes, and TBI in military vets, Faden argues, may have inadvertently "minimized the fact that concussions happen far more often in the normal population and in the elderly" than we've realized. There are nearly four million head injuries per year in the United States, he emphasizes. "That's a very high number, one that's not well appreciated." Think of middle schoolers playing soccer and lacrosse, kids who fall off their bikes and skateboards, parents who tumble off the roof while fixing the gutters or lose their balance on the stepladder while hanging holiday decorations.

Such everyday concussions can significantly alter a patient's life. Faden has shown that when there is a traumatic insult to the brain or spinal cord, "it generates inflammation in the brain that can occur for months or years and lead to progressive loss of brain cells, as well as continued tissue destruction," he says. As Faden talks, he moves his hand, palm down, back and forth through the air above his crossed legs in a

---

* Historically, the loss of cognitive function following repeated blows to the head was first described in boxers as early as the 1920s and was referred to as *dementia pugilistica*.

thoughtful, almost professorial gesture, as if to make the delivery of these discoveries a little less terrifying.

Other researchers who have been studying concussion have arrived at similar conclusions. A single "moderate" traumatic brain injury has been shown to cause inflammation to continue to brew in the brain, contributing to cognitive decline, depression, mood changes, and memory loss for years down the line. Children, girls, and women with brain injuries are more likely to develop anxiety, panic attacks, and depressive disorders even thirteen years later (that's as far as researchers have followed these women). The brains of those who've suffered concussions also appear to be five years older than their chronological age. In one recent study of 235,000 patients' health records, individuals who suffered a single mild concussion were three times more likely to commit suicide, even nine years after their initial injury. And, says Faden, more than 40 percent of those with even a single concussion still show significant disabilities four years later.

Faden has proposed the term "chronic traumatic brain inflammation," or CTBI, to distinguish this more common condition resulting from head trauma from CTE, with an emphasis on the fact that CTBI, unlike CTE, is a treatable disease.

Brain Injury and Frenzied Microglia

So how can it be that a blow to the head can wreak such havoc to the nimble workings of the mind, and continue to do so year after year, long after the injury has occurred?

Researchers have known for some time, based on autopsies of the brains of individuals who suffered traumatic brain injury and who died years later from other causes, that these brains showed an unusual level of microglial activity. But this frenzy of agitated microglia was not seen, says Faden, as an important factor in progressive brain damage.

Faden and his team began looking more closely. Using MRI imaging, they examined the brains of mice who'd been exposed to moderate traumatic brain injuries and found that microglia remained overexcited for up to a year following the original injury. These mice also showed neurodegeneration in the brain's hippocampus, and their biomarkers for neuroinflammation were significantly higher than normal.

"We wondered what part microglia were playing in this and found, to our great surprise, that they were taking on myriad roles. We could distinguish many different populations of microglial cells," he explains. "The microglia weren't just either good or bad, yin or yang. There were many different subsets with different neurotoxic capacities." Many of which caused mayhem in the brain.

And once researchers like Faden began to take these newly understood immune cells in the brain into account, the field's understanding of concussion catapulted forward.

Faden's investigation into microglia's role in concussion took place just a few years after Beth Stevens did her groundbreaking work at Harvard showing that some triggered microglia were eating synapses, while others were spewing out inflammatory cocktails that caused neuroinflammation.

Faden now believes that the activation of toxic microglia is the major contributor to chronic traumatic brain inflammation, and the associated brain cell loss and brain dysfunction that is the hallmark of CTBI. An injury to the head triggers microglia to switch from protecting and repairing the brain to excreting inflammatory chemicals that create more microglia-driven runaway inflammation. When this happens, activated microglia increase in size, becoming big and bulky and bushy; they look like loping tarantulas under the microscope. They bite away at synapses, leading to loss of memory, concentration, clarity of mind, and buoyancy of mood.

Once these microglia forest fires get going, without intervention, they become harder and harder to put out, and CTBI can develop.

But because scientists missed, for so long, the fact that microglia could become chronically active and destructive after head injury, clinicians weren't making this crucial connection and utilizing it to improve patient treatment—not by a long shot. For instance, if a boy was knocked out cold by a soccer ball in middle school and then developed depression or a panic disorder or struggled academically in high school, the link between his earlier head injury and later learning issues and depression flew completely under the radar.

Which leads us to Faden's most recent—and, to my mind, most surprising—discovery. In 2017, Faden found in animal studies that those with a traumatic brain injury had higher levels of unique microparticles in the blood, as compared to other animals. When Faden traced where those particles originated, he found that they were being directly released by microglia. After an injury, the response of microglia was so out of proportion that microglia were driving these microparticles into other areas of the brain, far from the site of the original injury, stoking up more inflammation, leading to more tissue damage. Astonishingly, some of these particles that microglia projected forth were getting released right into the body's bloodstream.

And, as scary as it sounds, his knowledge is very helpful and promising news.

Remember Beth Stevens's hope that we will soon possess clear blood biomarkers to accurately measure factors that microglia secrete as they begin to shift from angels to assassins, so that we can better grasp what microglia are doing in the brain, in terms of denuding synapses, through simple blood tests?

Once these microglia-derived microparticles can be accurately measured by a common blood test, these biomarkers can be used to monitor concussion treatment and healing. Imagine, for instance, that a woman has a mild head trauma in a car accident. Clinicians may one day be able to test her blood to see how much inflammation is brewing in her brain and how severe her concussion is, and then continue to run routine blood tests to evaluate how well she's responding to treatment. If lab tests show she still has high levels of microglia-driven inflammatory microparticles in her blood, she will need further treatment and observation. If levels of microglia microparticles rapidly decline, practitioners will know she's well on the road to healing.

Faden's lab has revealed one other remarkable finding. Traditionally, he says, research on spinal cord injuries has neglected the effects of those injuries on brain function. But spinal cord injuries, he says, can cause widespread and sustained brain inflammation, progressive loss of brain cells, cognitive decline, and depressive symptoms—by provoking microglia to create invisible havoc in the brain.

This bidirectional chatter between the central nervous system's immune response (via the cerebrospinal fluid, intersecting with the lymphatic pathways that run through the meningeal spaces of the brain) and microglia makes it even clearer that brain and body are in constant conversation.

For instance, in one 2017 study, Swedish epidemiologists at the Karolinska Institute scoured eighty thousand adult health records and reported that teens who'd experienced a single concussion were 22 percent more likely to later develop multiple sclerosis, compared to those who'd never had a head trauma—and for those with multiple concussions, the increased risk of developing MS rose to 150 percent. At this point, given what we know about the feedback loop between body and brain—and the way in which microglial cells in our brain and the immune cells in our body chitchat—this is hardly surprising.

So let's pause for a moment and acknowledge that all of this new information about head injury and concussion is pretty frightening. It certainly scares me. In addition to my autoimmune issues, I've had two mild concussions. Twenty-five years ago my husband and I were going to the movies. When we got to the movie theater, I dashed out of the car to buy the tickets while he parked. My purse strap had become wrapped around the stick shift, and as he pulled away, my head ricocheted back, hitting the roof of the car. Back then, they called it a brain bruise. The second time, sixteen years ago, I was a passenger in a car driven by a friend. She hit a patch of black ice while we were going down a hill and we hit a telephone pole—and I struck my head on the passenger side window. I rested, iced, recovered, little by little. But it makes me wonder. The same year that I had that second concussion in the car accident, I later developed Guillain-Barré syndrome—a disease similar in pathogenesis to multiple sclerosis—for the first time. Might there have been some connection? It is impossible to say. But clearly, the idea that chronic traumatic brain inflammation may be occurring and causing unsuspected mental and physical havoc is dismaying for any individual, or parent of a child, who's had a head injury.

I tell Faden that I'm afraid that all this information is going to be very scary for readers—they'll want to throw this book right out the window. "I imagine it does sound frightening," he says. But then his face brightens. "The *most* important message, though, is this. Although CTE is not a treatable disease, mild and moderate concussion are on the cusp of becoming *very* treatable." As he talks, he gently punches the first two fingers of his left hand into the air with excitement. For decades Faden has been working to develop anti-inflammatories that have the potential to ameliorate and reverse brain injury, alongside other interventions— and he feels that new approaches may be able to limit bad outcomes after concussive brain injuries.

Over the course of his career—long before there was an understanding of the role that microglia play in concussion—Faden and his colleagues went through a cycle of hope and frustration, introducing an array of different drug treatments into the bloodstream—including selected anticancer drugs, thyrotropin-releasing prolactin (or TRH), and glutamate-blocking drugs. Delivered within a few minutes, hours, or, in some cases, days of injury, these varied therapies dramatically reduced the level of neurotoxicity and cell death after a head injury.

But the problem has been, Faden says, resting his palm in his cheek for a moment as he continues, that because some of these agents were no longer subject to patent restrictions, pharmaceutical companies weren't interested in investing in them. There was no money to be made. Unfortunately, says Faden, "Despite the fact that these medications show exceptional promise experimentally, they will never be examined in large clinical trials."

So what new treatments are on the horizon?

Once again, it all comes down to the promise and peril of microglia. Faden has found that by using a combination of newly studied and noninvasive approaches, physicians can help patients tamp down overexcited microglia, even weeks or months after a mild or moderate brain injury has occurred—making it possible to intervene to help a far wider swath of patients more of the time.

Right now, Faden's lab is studying the effects of combining "three simple and easily available approaches," he explains. "These include aerobic exercise, dietary manipulations such as an intermittent fasting, and computer brain training."

UCLA researchers recently found that exercise—done within certain windows of time after a concussion and only with a doctor's approval—increases chemical factors in the brain that damp down the hyperactivity of microglia. Faden's lab is also the first to investigate whether an intermittent fasting diet—in which patients fast for long stretches in between normal eating—will increase protective brain factors when done in conjunction with aerobic exercise. (Animal studies on intermittent fasting after traumatic brain injury have been promising.) He hopes that when they also add in computer brain training to improve cognitive function, alongside exercise and dietary changes, they will see added benefits.* Right now he and his colleagues are testing this simple triad of ideas on mice, but they plan to conduct human trials in the future.

Together, these new investigations will give us far more insight into how, says Faden, "we can potentially intervene months or even years after an injury, offering novel treatments long after the initiating insult."

Faden and others are now experimenting with delivering a new microglia-targeted treatment that can eliminate almost all microglia—both the good guys and the bad guys. After treatment, newborn microglia begin to repopulate the brain, and as they do, they become neuroprotective rather than neurotoxic. The big bad assassin microglia are no longer on the scene; only the angels are perusing the brain. One month later, "the toxic population of inflammatory microglial cells is still markedly reduced," Faden, says, his voice full of excitement.

It's a little like rebooting your computer—hitting the coding keys to

---

* Patients with concussion or head injury should always consult their physician or concussion specialist for advice about when and if it's appropriate, in their particular case, to consider exercise, a fasting diet, brain training, or neurofeedback. None of these therapies should be undertaken other than at a physician's suggestion and under his or her supervision.

delete everything and rebuilding your operating system over again without the computer viruses or glitches.

This oral anti-inflammatory medication—what we might think of as a kind of fire extinguisher for the brain—is still several years away from clinical trials, let alone being available to patients, but it adds to the sense of hope on the horizon for patients with concussions and brain injuries.*

"If we can change the level of microglia-driven inflammation through these multiple interventions simultaneously, we should be able to change patients' outcomes after brain injury," says Faden. "We are getting so much closer."

---

* Of course, like all new drug therapies, this treatment will need to go through extensive clinical trials that will have to be replicated and repeated in order to evaluate efficacy, safety, and potential side effects.

FOURTEEN

The Fast-er Cure?

LILA SHEN AND I ARE STARING DOWN AT THE BEAUTIFULLY PACK-aged white box that sits on her round pine breakfast table. It has just arrived on her doorstep, courtesy of UPS. The words PROLON: PROMOT-ING HEALTH AND LONGEVITY are printed in elegant green letters across the top. Lila is about to try ProLon, which is a fasting-mimicking diet developed by researchers at the University of Southern California to help boost immune health and brain function. Lila's primary motiva-tion, she says, is the increasing brain fuzz and memory glitches she faces, especially during flare-ups of her inflammatory bowel disorder, Crohn's disease.

Lila (whom we first met in chapter 4) gives me a few examples of her latest mental "brain farts," as she calls them. "It's beyond opening the fridge and forgetting what I wanted to get out." She sighs, exhaling slowly from between pursed, worried lips. "We all do that sometimes. My glitches border on being dangerous. I've burned up pans and teakettles on the stove so often that my husband bought me an electric hot water machine, a rice cooker, and a crockpot. Whenever I'm cooking I have to set a timer to remind myself that I've put something on the burner before I turn around to do the next thing, just so I don't forget, walk out of the room, and burn down the house. I've missed doctor's appointments, forgotten to put the car in park before I turn it off, and left the dog in the yard and driven off for the airport!" At work, where Lila manages a fund-raising office for a small Washington, D.C., nonprofit, she has two white-boards with color-coded Post-it notes to help her keep track of who is doing what, where, and when. Even so, she has been letting the ball drop more than she used to, and this makes her nervous.

Her reputation as a rainmaker, bringing in donations, is such that her job is not in jeopardy, but she does find herself wondering, "How long can I stay in this position if I can't remember who donated what, or what their names are, when I see them at events?" She confides to me that she sometimes thinks of the TV show *Veep*, in which the main character, a scattered female politician, has an assistant whisper in her ear to remind her who she's schmoozing with and what she should remember about them as she works her way across the room. "My staff has had to step in and save me a lot more frequently lately," she says. "It's unsettling."

Add to this normal aging of both body and brain, and, well, the bottom line is that Lila feels as if she is barely muddling along. Focusing, staying on task, recalling things she needs to recall at will—feels (and has felt for a while now) a bit like "trying to peer through a window to see the outline of the tree leaves and the clouds, but the glass is smeared with Vaseline."

It seems all too clear that the inflammatory bowel autoimmune disorder Lila faces may have also affected her cognition, mood, and clarity of mind—as is sometimes the case for patients with Crohn's.

Indeed, the microglial universal theory of disease tells us that this is exactly what is likely to be going on in patients like Lila. Inflammatory processes in the body are—in a complex cascade of neurobiological events—chitchatting away with the agitated microglial immune cells in the brain, sparking bidirectional neuroinflammation that can further kindle brain fog, memory short-outs, obsessive-compulsive worries, and sudden upwellings of sadness.

Of course, if Lila's body's overexcited macrophage immune cells and her brain's microglia are conversing 24-7, everything that she can do to help calm her mind and restore a sense of mental equilibrium will help her brain and body too. Hundreds of research studies tell us that strategies to tame stressful thoughts and rumination also help to reduce the day in, day out toxic cocktail of "fight, flight, or freeze" inflammatory chemicals sent forth by our nervous system whenever our stress response is set on high. Lila makes good use of a number of these mind-body methods. She is a fan of body scans, guided imagery, breath work, neural "self-talk," hypnosis, strength training exercise, restorative yoga, and walks in nature, all in an effort to downshift her stress response and calm down her immune system in any way she can.

But for patients like Lila—especially as they get older—simple life hacks like yoga and meditation may not be enough. Especially if the body's hyperactive macrophages and other immune cells keep shouting up to the brain's microglia, "Hey, you guys! We have a problem down here! Get ready to take extreme measures!" Which in turn can signal microglia to keep misbehaving, inflaming and destroying synapses, leading to more forgetful, fuzzy, funky brain glitches—and pans going up in flames on the stove.

That's no good at all.

So Lila has been wondering what more she can do about her own overbusy microglia. She's a good candidate for neurofeedback and TMS, but at this moment, her insurance doesn't reimburse patients very well for these treatments (although many insurance plans do), and she is already inundated with medical expenses.

Is there a way for someone like Lila to target both physical and neuro-inflammation at the same time, in more of a do-it-yourself way?

Enter the newest and perhaps most gloriously simple DIY brain hack scientists are offering up to better address physical and neuroinflammation simultaneously: the fasting-mimicking diet. At the University of Southern California's Longevity Institute, director Valter Longo, Ph.D., has spent the last two decades looking at the role of structured, science-based, safe fasting in immune health and overall longevity.*

* None of these fasting dietary approaches should be undertaken without the advice and guidance of your physician. It's also enormously important to underscore here—

Longo, now in his late forties, appears a good ten years younger than he is. His dark hair, parted boyishly in the middle, just brushes his shirt collar. I ask him to help me connect the dots between the fasting-mimicking diet, inflammation, and microglial-driven neuroinflammation. (Since Longo is on the West Coast and I'm on the East, we chat by phone and Skype.)

In his early days as a young scientist at UCLA, Longo tells me, he began experimenting with yeast organisms and discovered that those organisms that had specific mutations in their growth genes were able to live up to five times longer than normal yeast. He began examining studies others had already done on similar growth gene mutations in mice, and he found that in some mice, these same genetic changes in growth genes were likewise linked to greater longevity. It appeared that in these mice a severely reduced level of growth factors, for reasons unknown, caused molecular changes that led to a dramatically increased life span. Just like yeast, the mice were able to enter a kind of "higher protection mode that allowed them to live longer," says Longo.

Longo began to look for similar gene alterations in humans, and he found that a community of people in the Ecuadorian Andes possessed a mutation in the same growth genes that give mice record longevity and health. These individuals lacked the receptor for growth hormone. This mutation also protected these individuals from cancer, diabetes, and age-dependent cognitive decline. In fact, their brains appeared much younger than they were; they had greater cognitive function compared to their relatives and others who were of the same chronological age.*

Even though this group of Ecuadorians ate a regular diet, mutations in their genes tricked their bodies into thinking they were starving. "It appeared that these individuals were, in a way, stuck in starvation mode,

---

especially given skyrocketing rates of eating disorders in adolescents and young adults—that we are not talking about disordered eating.

* The endocrinologist Jaime Guevara-Aguirre first studied this community in Ecuador, a group of extremely short individuals who lack the receptor for growth hormone in a disorder known as Laron syndrome.

even though they were eating normally," says Longo. And this helped to protect their bodies and brains against the effects of aging.

These findings, while striking, made intrinsic sense to Longo. "We have known for a long time that when we starve bacteria and other microorganisms, we can get rid of all the junk," he explains enthusiastically in his faint Italian accent. "In mice, we've also discovered that if the liver or pancreas is damaged, fasting helps to rebuild it." So, Longo wondered, "Could periods of intermittent fasting also help to activate longevity genes in the general population in a way that would help get rid of damage not only in the liver, but also in the body's entire immune system?"

To find out, Longo placed one group of mice with an autoimmune disease similar to multiple sclerosis (these mice did *not* possess the longevity growth gene mutation) on a fasting-mimicking diet for three days once a week for three weeks. A second group of mice were fed a normal diet. Mice on the fasting-mimicking diet showed a marked reduction in inflammation-causing cytokines. But more astonishingly, the diet promoted a regeneration of myelin—the coating of proteins and fats that wraps around and insulates the nerve fibers that run between the spine and the brain and that the body's immune cells attack in neurological diseases like multiple sclerosis and Guillain-Barré. Longo published his groundbreaking findings in 2016.

"When you have a person fast in a safe, controlled, and measured way, everything in the body shrinks down a little, including the immune system," says Longo. "By tricking the body into thinking that it's running out of fuel, the immune system uses this opportunity to reduce its function, refresh, and restart; it gets busy destroying damaged autoimmune cells. When you begin to add calories back in, everything rebuilds—only this time, at least in mice, the damaged cells are gone and healthy new stem cells repopulate the body. They start making necessary repairs."

Degenerated myelin regenerates.

In animals on the fasting-mimicking diet (as compared to a control group), there is a noted reduction of both inflammatory macrophages in the body and inflammatory microglia in the brain. Both body and brain hit the Refresh button.

Meanwhile, another leader in this growing niche of science—now known as *biogerontology*—Mark Mattson, Ph.D., chief of the Laboratory of Neurosciences at the National Institute on Aging and a professor of neuroscience at Johns Hopkins University, has showed that an intermittent fasting diet increases the resistance of neurons in the brain to synaptic pruning and inflammation in animal models of Alzheimer's disease, Parkinson's disease, Huntington's disease, and stroke.* In one 2018 study, Mattson demonstrated that fasting improved cognition and mood by suppressing microglia-mediated infection, thus helping to protect neurons against stress and allowing for neurogenesis.

Longo felt—and has staked his scientific career on the idea—that the fasting-mimicking diet (known for short as the FMD) might turn on "nature's oldest and most direct fix for an ailing immune system" and that careful fasting could kick-start "a regenerative and self-healing process, with minimal or potentially no side effects."

But mice are not men. Or women. Longo wasn't sure whether the same thing would happen in clinical human trials. He knew that the trick in working with people would be to develop an FMD that would provide sufficient calories so that it could be used safely outside a clinic setting, even at home. Longo wanted to reduce caloric intake enough to convince the body it was fasting, while ensuring that a patient was getting all the nutrients and vitamins needed to support rejuvenation. And he wanted to provide food that, well, just tasted good. The bottom line: all the health benefits of fasting, with all the essential nutrients and without any fainting episodes, hunger pangs, or sense of deprivation.

He scoured data on how specific nutrients influence the function of

---

* The fasting-mimicking diet is similar to two other types of fasting, the intermittent-fasting diet and time-restricted fasting, approaches that Alan Faden, M.D., referred to in chapter 13 as part of a novel combination approach to help treat brain injury. Intermittent fasting programs include what's referred to as the 5:2 program, in which an individual eats normal amounts of nutritious food five days a week, and two days a week eats only one moderate meal limited to 500 to 600 calories. Time-restricted fasting refers to fasting on a schedule—for instance, an individual eats during an eight-hour window during the day but doesn't eat for a longer window of time such as sixteen hours (for example, fasting between 7:00 P.M. and 11:00 A.M.).

immune cells, affecting a number of autoimmune and brain-related disorders, and studied diets from around the globe associated with longevity and strong cognitive function. He traveled to observe populations that enjoyed lower rates of illness and cognitive decline. For instance, the aging population of Okinawa, Japan, showed strong benefits from their fish-and-plant-based diet: less cancer and heart disease as compared to Americans of the same age, and a 50 percent lower rate of dementia. Longo noted similar correlations when he visited other geographical zones where the elderly enjoyed surprising longevity and brain wellness — including in his own Italian grandparents' birthplaces of Sardinia and Calabria, where people consumed a mostly plant-based Mediterranean diet.

He began testing the fasting-mimicking diet in humans at the University of Southern California Medical Center. One hundred participants enrolled in his study. These individuals ate a special low-calorie, nutrient-dense, plant-based diet, painstakingly designed by Longo, for a period of five days a month for three months.

The results were extraordinarily promising. Longo found that patients who did the FMD showed a range of positive biological changes: increased muscle mass relative to their body weight, decreased blood pressure, and lowered cholesterol. In patients who were prediabetic, blood glucose levels returned to normal ranges. Patients also reported improved learning ability, memory, greater clarity of mind, and more energy.

The fasting-mimicking diet appeared to work in the same way in humans as it had in animal studies, says Longo, possibly by breaking down and regenerating the inside of cells, killing off and replacing damaged cells, and elevating levels of stem cells circulating in the blood so that they could start repairing, rebuilding, and regenerating.

Over the past several years, more research on fasting has come out, showing that it reduces obesity and risk factors for diabetes, cancer, and cardiovascular and neurodegenerative disorders.

After much scientific refining and clinical testing at the University of Southern California's Longevity Institute and School of Medicine, Longo's fasting-mimicking diet is now available for patients to try at home.

. . .

"What do you think is happening to microglia in the brain, in specific, during the fasting-mimicking diet?" I ask Longo. "How is fasting calling off rogue microglia, helping good microglia to support neurons again, and, to some degree, reversing neural and synaptic damage?"

"We know that an excess of microglial activity can damage cells in the nervous system," says Longo. "And we know that the fasting-mimicking diet is having an effect on microglia and microglia-led inflammation in the brain, because we can see that the FMD decreases inflammation in the brain and improves cognition in mice." He pauses. "But as to the exact biological pathways by which the FMD is changing the activity of microglia, we are still working out those specifics."

Right now, Longo is "particularly interested in fasting in the context of Alzheimer's disease" and is starting a clinical trial of the FMD in Alzheimer's patients. He also believes there is "the potential for even more powerful health benefits for patients with chronic conditions by combining the FMD with specific drugs." Currently, he's investigating that hypothesis in human clinical trials of patients with multiple sclerosis in Italy, pairing the FMD with each of five different drugs that target the immune system in different ways. "After we target the damaging immune cells with drugs, we add in the FMD, in order to repair everything and make cells come back in a healthy way."

A drug treatment alone, Longo underscores, "is not going to bring healthy new cells back into play." But the fasting-mimicking diet can be utilized as the second step in a two-punch process—for instance, using cancer agents to destroy cancer cells, then employing the fasting-mimicking diet to enhance the destruction of cancer cells and to help restore and rejuvenate healthy cells. In one stunning finding, Longo recently discovered that many types of cancer cells—including those in melanoma, breast cancer, and colon cancer—that are resistant to conventional treatment are more likely to die off during periodic fasting cycles.

A Gut Feeling: The Microglia-to-Microbiome Connection

Another way of understanding this diet-brain interaction is to consider what we know about the strong connection between gut health and brain health. Stacks of studies tell us that unhealthy alterations in the gut's microbiota can trigger depression and other mood disorders in the brain. In one study of women between the ages of eighteen and fifty-six, those suffering from major depressive disorder had a significantly different constellation of gut microbes—higher levels of sixteen different types of gut bacteria—compared to those without depression. The makeup of bacteria in the gut microbiome also appears to be different in patients with mental and physical illnesses including anxiety disorders, obesity, anorexia, Parkinson's disease, and multiple sclerosis.

Chronic stress, which also promotes an inflammatory immune response, can change the composition of the gut microbiome, increasing populations of harmful microbiota, which in turn affects gene expression, increasing inflammation in the central nervous system and influencing mood and behavior in a never-ending feedback loop between gut, body, and brain.* Our microbiome also directly affects levels of neurotransmitters like serotonin in our brain, and can even directly produce neurotransmitters, the chemical messengers that convey signals between synapses in our brain and nervous system. This general understanding that microbes in the gut communicate with the central nervous system and brain through neural pathways is often referred to as the "gut-brain axis."

One 2018 study demonstrates the potential power of addressing the gut-brain axis in patient healing. Researchers at Johns Hopkins used probiotics—comprising several healthy gut microbes—in psychiatric patients to see if doing so might improve both their immune function and

---

* Recently, scientists at the University of California, San Francisco, were able to create what they called a 3-D mini-gut. They found that receptors on cells that line the gut are able to recognize stress hormones, such as adrenaline, when the brain sends them cascading through the body. Once these gut cells receive stress hormone signals, they begin to interact with nerve fibers, producing compounds that are similar to those that are made when synaptic connections light up in the brain.

psychiatric symptoms. Scientists focused on patients with bipolar disorder who had a history of being hospitalized for mania—since, as we've seen, many patients with bipolar depression have elevated levels of inflammatory cytokines during manic episodes. Patients with bipolar disorder who took the probiotic along with their usual meds for six weeks were 75 percent less likely to be readmitted to the hospital for manic episodes compared to those who didn't.

This may be the case because changes in signaling from the gut microbiome dramatically influence the activity of microglia in the brain too. When the gut microbiome is triggered by changes in the environment—stress, injury, a low-nutrient, highly processed food diet—immune cells in the gut interact with macrophages that can spit out inflammatory cytokines, leading to higher levels of pro-inflammatory chemicals speeding along the brain-immune highway. Microglia, in response, grow in size and number in the brain—and more synapses are pruned. When the gut microbiome is comprised of a population of gut bacteria that's less inflammatory, microglia in the brain may become less inflammatory too.

Two brains—the brain and its central nervous system, and the "second brain," our gut brain—talk, argue, negotiate, and send and decode messages back and forth in a kind of never-ending Morse code—and when these signals are received, they can either calm microglia or rev them up to do more harm.

With fasting, we may have some ability to influence microglia and help these tiny cells turn from the dark side to the light again.

So, if our gut microbiome (in simplest terms) plays a significant role in influencing microglia to either repair or snip synapses, and a fasting diet can direct the gut microbiome and microglia in positive ways—well, that seems like a win-win all around.

Study of One

Since ProLon is a medical dietary protocol, the company Longo started, L-Nutra, requires patients like Lila to complete a full medical intake form, which is then reviewed by a nurse. If the patient is in excellent health, she or he will able to purchase the dietary kit on her own through the website.*

If the patient is not in excellent health, she or he will need a doctor's prescription. This is for a very important reason. "We would like to wait for the results of several clinical trials before recommending that patients with autoimmune disease try the fasting-mimicking diet at home," says Longo.

Lila calls her internist for his advice. He does a little investigating and gets back to her, telling her he's looked at what foods the FMD includes and the dietary regimen involved, and he is okay with Lila trying it. He goes on the ProLon website and writes Lila an online prescription for the fasting-mimicking diet. But, he says, he has two conditions. Lila should do the fasting-mimicking diet during the week, so that if she encounters any issues—abdominal pain, diarrhea, faintness, dizziness—he will be able to see her right away. And if she develops any symptoms that are worrisome—anything more than her normal ups and downs—she is to stop immediately.

Lila feels that since she suffers from both gut and brain-related disorders, and her doctor has given her the okay and will be standing by, she wants to at least give it a try.†

---

* It's not necessary to purchase a particular dietary program to see if a fasting diet protocol can benefit your health. Research points to similar health benefits from intermittent fasting diets that you can try (under the supervision of your doctor) at home. These include intermittent fasting (limiting caloric intake to 500 calories on two nonconsecutive days a week) and time-restricted fasting (consuming all of your food for the day during an eight- or ten-hour period, then fasting for fourteen or sixteen hours overnight). Likewise, says Mattson, "If you consume only seven hundred calories—in healthy foods—for five days in a row each month, that will result in metabolic changes, which will lead to the same improvements in health and cognition." Ask your doctor if intermittent fasting is appropriate for your health situation, and request specific guidelines before beginning any protocol.

† The fact that Lila (while under her physician's supervision) opts to try the FMD is not intended to encourage any reader with an autoimmune or other condition to try the FMD. The literature that comes with ProLon states, "ProLon is *not* intended for the following people: People diagnosed and currently suffering from a medical condition or disease. People that have been severely weakened by a disease or medical procedure. People with a history of syncope (fainting)," and so on. The literature also lists possible side effects, including fainting and dizziness as well as spinal pain. Patients

She has ordered one kit and plans to begin it at the start of the next five-day workweek.

And so here we are, unpacking Lila's box in her kitchen. The ProLon package is elegantly put together—albeit a little alarming in that each day's food, which is separated into its own white "Day Box," comes in a package that is only about twice the size of that in which a new iPhone might arrive. Each day's fare consists of some combination of Longo's carefully designed plant-based meals: prepackaged, nonperishable, low-calorie, and nutrient-dense soups, bars, drinks, and snacks, along with herbal teas, vitamins, and supplements—all to be eaten on a schedule.

"It's all so bare-bones," Lila says as she pulls out the small packaged food items from the box labeled Day 1. Breakfast is a macadamia, almond, coconut, and flaxseed bar called the L-bar, which is taken with omega-3 and DHA supplements (these both come in the form of a pill). Morning tea on the FMD is caffeine-free, spearmint-lemon-flavored. Lunch is a packet of tomato soup with sides of olives and a few kale crackers. (Olives appear to be a big part of this diet—and these olives are thankfully fabulous olives with garlic, harvested in the vineyards of Italy.) In the afternoon, Lila will take two small vegetable powder vitamin tablets called NR-1 and snack on a second L-bar. Dinner is a gluten-free minestrone soup topped off with a small chocolate cookie bar called the Choco Crisp.

"I feel as if I'm going on a spaceman diet or something," Lila quips.

I laugh. Yes, that's exactly what it looks like: astronaut food.

"But if it helps . . ." she says, giving a minuscule shrug as she weighs the breakfast bar in her palm. "If this tells the little microbes in my gut to tell those glial cells in my brain to back off and behave, hey, I want to give it a try."

---

should consult their doctors before starting a fasting diet. As with all modalities discussed in this book, this is not intended to be a substitute for professional medical advice, diagnosis, or treatment. Always seek the advice of your physician or other qualified health provider with any questions you may have regarding a medical condition or treatment.

Before Lila starts the diet, as a kind of self-test, she sits down at her computer and takes an online "word list recall" memory test similar to those given to patients in neurology clinics to test short-term working memory. I sit beside her as she tries to memorize fifteen different words—each of which is displayed on her computer screen for a few seconds. Then, after all the words have appeared and disappeared, Lila writes down as many she can remember. She recalls five words. Disappointed, she plays the game again. Four words. Again. Six words.

Day One

The Day 1 protocol on ProLon consists of 1,200 calories, and as such, it offers more food than the rest of the days on the five-day plan. "Honestly, it's not that much less than what I would usually eat on a normal day, or at least that's what I'm telling myself," Lila says.

Around seven thirty that night, I call her and ask, "Do you feel hungry?"

"No."

"Anything different at all?"

"I feel pretty nourished," Lila says. "Just a normal day, really. I forgot to pick the dog up from the groomer on my way home from work and only remembered when I walked in the door and saw she wasn't here to greet me. I had to run back out to get her." She sighs. "Normal day!"

Day Two

On the second day, Lila follows the instructions that come with the Pro-Lon kit and stirs a proprietary potassium mixture, called the L-drink, into a half gallon of water so that she can sip on it all day. The kit comes with an extra-large plastic water bottle that sports the ProLon logo on the side.

"I'll be sipping on this in all of my meetings," she says. Adding in the L-drink is not the only difference between Day 1 and Day 2. "There is less food on offer today," she jokes. "No kale crackers. And I loved the kale crackers! And no afternoon snack bar, either. But I'm committed. One day down, four to go."

The ProLon team has sent Lila an automatic email telling her what to expect on Day 2. It offers a soft caution: Many people feel fatigued on the second day, so take it easy, don't do too much. "Fatigue is normal as your body is starting its transformation. Remember to not overexert yourself today."

This prediction turns out to be all too true for Lila.

She has two meetings that day—one in her office and one later in the afternoon across town in Washington, D.C. "By the time I got back to the parking garage after my last meeting, I was just dragging my feet along," she tells me. "I was walking up the stairs in the garage when a wave of fatigue hit me like a tsunami. It was hard to get up a single flight of steps. My head was pounding in my skull. I felt like something had sucked all the blood out of my body. It was like I'd come down with the flu."

Still, she stayed with it. "If this is the worst of it, I can gut through it." She laughs. "Ha! No pun intended!"

"Dinner was another tiny space meal," Lila also tells me, by text, later that evening. "But all I can think about right now is crawling into bed!"

Day Three

The next day I check in with Lila by text.

"Similar to day 2," she writes. "Tired! But maybe less so? Fell into bed at nine o'clock and woke up nine hours later. I can't think of the last time I slept nine hours!"

"How is your brain working?"

"Maybe something is happening," she texts. "More grounded. Alertish. AWAKE. Brain less out of control? Brain is buzzing in a good way instead of buzzing in a bad way? Not sure yet."

We agree to talk again tomorrow.

Day Four

"Starving!" Lila texts me the next morning at 7 A.M. "Woke up 4 A.M. so hungry I wanted to eat my pillow! Too hungry to sleep!" Then she adds, "But brain feels lighter. CLEARER."

I phone Lila that evening to check in. She's just finished feasting on reconstituted quinoa soup with olives on the side, and she catches me up on the last forty-eight hours. "I was running this big meeting and suddenly I realized that I was able to visualize some of the moving parts for one of our upcoming events more clearly—I could hold most of the pieces in my mind at the same time—in a way that has, of late, been hard for me to do."

After the meeting, Lila says, she got out all of the seating place cards for the upcoming gala, spread them out on the conference table, and re-solved a seating chart nightmare—something her staff had been fretting about all day—so that all donors would feel happy and valued. "This feels a little bit more like the *old* me—the one that could see patterns, solve problems other people were struggling to solve. That's why I got into this business. I like seeing how puzzle pieces go together in ways that can make everyone happy and do good for the world at the same time."

Day Five

It's Lila's last day on the fasting-mimicking diet. Today her diet consists of a morning L-bar, tomato soup, kale crackers, and minestrone soup.

"I'm not as hungry today," she says. "Maybe my body is adjusting?"

She fills me in on a few other subtle changes she's been noticing. "It's easier to drive, for one thing. Usually when I'm pulling out of a parking lot, I'm really struggling to figure out where other cars are around me in space, and I'm worried about hitting something. I've had so many little fender-benders in the past four or five years. But after work, I pulled in and out of a few parking lots while running errands and I realized afterwards that I'd done it so much more smoothly than usual. It was a piece of cake.

"I feel more in my body, more alert, more clear-headed, less aggrieved, less worried about my future . . . more *me*," she adds.

The real test, she says, will come the following evening, at her organization's annual gala in D.C., where she'll need to remember every donor and his or her spouse's name, where his or her kids go to college, and so on, so that she can converse with them in a way that makes them know how much their generosity is appreciated.

"Will it be hard not to eat the mini crab cakes and sip champagne?" I ask.

"No!" she says. "None of that really even appeals to me right now. Besides, I never have time to eat at these events, and I never drink at them, I just pretend to sip the champagne. I need every little microglial cell I have to be working for me."

Day Six

On Day 6, although Lila's five-day FMD is officially over, the kit's instructions suggest that she eat lightly. It is a transitional day. She has toast for breakfast and soup for lunch, and she is surprised to find, she says, "I'm just not that hungry."

She gets dressed for her gala, taking note, she tells me, "of a little less middle-aged spread around my middle." Her dress is almost loose, she says as she heads out the door with her husband.

Day Seven

On Day 7, Lila and I go for a walk together in the park. Lila is back to her normal diet. She's just finished making scrambled eggs and toast for her two boys and eggs for herself.

"I'm not as ravenous as I thought I'd be," she tells me. "Oh! And I didn't have to set a timer while I was scrambling eggs! I wasn't worried that I'd forget that I'd started making them, turn around to fill the dog's water bowl, and then walk off and burn everything. That scattered feeling has sort of dissipated. Not completely. I'm still easily overwhelmed when there is a lot going on around me. But my brain feels like it's had a good rinsing out."

"How is your stomach?" I ask.

"Maybe it's my imagination, but my gut feels a little less reactive," she says. "I'm not having as much abdominal pain. Less gas! Also good for work meetings." She laughs, then says, more soberly, "Usually I am thinking about my gut all the time because it's bothering me so much. When I'm cooking for the family, it's second nature for me to ask myself, *Can I eat this right now?* But I just realized, I'm *not* thinking about how my gut feels, which must be a sign that my gut is not as unhappy."

I share something with Lila that Japanese researchers recently reported finding while studying the benefits of fasting. They discovered that intermittent fasting, in animal studies, dramatically reduced inflammation in the colon. So, if we put together several recent findings from varied disciplines of science, the fact that Lila's gut feels better makes good sense. We know that patients with Crohn's disease are more likely to experience profound synaptic changes in the brain during flares when the colon is inflamed—caused by microglia-mediated inflammation activated in the brain's hippocampus. And this in turn can lead to more cognitive and mood disturbances. We also know that our gut's microbiome is always communicating with microglia. So it may be that the FMD is not only helping to alter Lila's gut microbiome in a good way, but, hypothetically, it's also having a soothing effect on the inflammation in Lila's colon, and on the microglia-brewed inflammation in her brain.

"So how was the gala?" I ask. "Were all of your donors happy?"

"I haven't really enjoyed one of these events in a while now, I've been too frantic about what might go wrong," she says. "And I've just felt so . . . tired." But, she says, "Last night, I felt like my old energy was kicking into gear. I'd see someone's face, and then their name would come flooding back to me. It was like I could suddenly retrieve old information from my mental files." She sighs a happy sigh. "It's such a *relief* to have a clear mind."

"Did you need your name-whisperer to help you remember who was who?" I ask.

"Not as much," she says. "I told her to stay nearby, but she rarely had to remind me of anything!"

Lila tells me that, out of curiosity, she plans to retake the fifteen-word memory test later that day, after she and her husband get their boys, Liam and Jason, where they need to go.

She texts that night. "I should be exhausted after past 48 hrs, but I'm not. Is that the FMD?"

"Might be!" I text. "Longo says patients have more energy after FMD. Did you do memory test?"

"Yes! Recalled 11 words!"

"WOW!" I text.

"If there were a drug I could take that would make me feel this clear? light? energized? happy? w/o side effects, would I?" Lila texts. "Yes, I WOULD."

I smile to myself.

"ALSO, MOST OF MY ECZEMA IS GONE" she texts. "And my nasal congestion." Chronic nasal congestion and skin rashes have plagued Lila for as long as I've known her.

And then, half an hour later, I get one more text from her: "Jason memorizing Frost poem. I'm helping him! Words of 'Road Not Taken' coming back in my mind from 4th grade!! ☺"

I am not sure, of course, how much of what Lila is experiencing is due to her synapses and microbiome both perking up with healthier cell populations, and how much is due to the energy the diet has given her, or even perhaps any ancillary placebo effect that accompanies her perceived gains. Still, I think it's fair to hypothesize that, in simplest terms, to some extent, Lila's microglia aren't beating up on her synapses in the same old way. Perhaps her synapses are even sparking and firing up new, healthy neural connections that in turn allow Lila to feel more like Lila again.

FIFTEEN

Future Medicine

A S THE MICROGLIAL UNIVERSAL THEORY OF DISEASE HAS BEGUN to be adopted by the scientific world, it is lending new credence to a range of already existing interventions that help to calm the behavior of overexcited microglia in the brain, as well as giving birth to new treatment approaches. Transcranial magnetic stimulation, neurofeedback, gamma light flicker therapy, novel concussion protocols, and fasting diets are just a few of the ways in which scientists hope to influence microglia and their interactions with neurons and synapses in the brain in order to help restore the quality of patients' lives.

Many new treatments on the horizon promise to take an extremely individualized approach to helping patients—one that is deeply informed by the microglial universal theory of disease, our new understanding of the bidirectional feedback loop between inflammation in the body and in the brain, and the role of neuroinflammation in treatment-resistant psychiatric disorders. Research is progressing at a rapid pace across multiple fields that may open up possibilities for personalized interventions guided by each patient's specific level of neuroinflammation and genetic profiles.

Where does such future hope lie?

Moving Beyond Serotonin

Certainly, the development of selective serotonin reuptake inhibitors (SSRIs) has been one of the most transformative new treatments for depression and anxiety in a generation. And indeed, the microglial universal theory of disease sheds new light on our understanding of the use of antidepressants—today's frontline treatment for depression and anxiety.

Three studies published in *The Lancet* and *Lancet Psychiatry* in April 2018 help us to better understand the complex interplay between microglia, depression severity, and treatment with antidepressants. In one study, researchers analyzed data from 522 trials and found that 21 common antidepressants were moderately more effective than placebo in treating adults with major depressive disorder. (This data was conflicting given that an earlier series of now famous studies had found antidepressants only a little more effective than sugar pills for treating depression.)*

A second study found that part of the reason for the moderate success of antidepressants appears to be due, yes, to the behavior of microglia. In two groups of patients with major depressive disorder, none of whom had ever taken any antidepressants, those who had suffered from untreated

---

* In an interesting aside, the so-called placebo effect is increasingly understood to be much more than a phenomenon in which people improve after receiving fake treatments (e.g. hand a patient a sugar pill and his headache vanishes for no discernible reason). Ted Kaptchuk, head of Harvard Medical School's Program in Placebo Studies and the Therapeutic Encounter, has shown that the placebo effect kicks in as a biological response to feeling cared for by one's physician or healer, and increases with the quality and quantity of the patient-healer relationship. Kaptchuk's colleague Kathryn Hall, Ph.D., a molecular biologist at Harvard, has found that individuals with certain genetic subtypes are more likely to experience an enhanced healing effect in response to a practitioner's act of caring. The gene that interests Hall is known as rs4680; it governs the production of enzymes that, in a series of complex processes, affect levels of brain chemicals including dopamine. Using fMRI scans, scientists can see brain areas associated with a heightened healing response light up more during healing encounters in the brains of individuals with these specific genetic subtypes of rs4680. What these findings suggest is that the psychological effect of placebo and the physical effect of a medication may influence healing along the same biochemical pathway. This pathway, which may be more likely to become activated in some individuals during caring medical encounters, generates powerful healing signals that turn on healing biological processes in both mind and body.

depression for ten years or longer showed greater levels of microglial activation and had a 33 percent higher level of proteins associated with neurodegeneration and the loss of gray matter volume in the brain, as compared to those in whom symptoms of depression had developed more recently. This suggested to researchers that the longer that depression went untreated, the more havoc microglia wreaked in the brain.

Then a third study, which looked at depressed patients who both had and had not taken antidepressants, showed something else quite staggering: This year-by-year increase in microglial activation and neurodegeneration was not as evident in patients who had taken antidepressants.

What does all this tell us? Not only does untreated depression do more damage to neurocircuitry over time, but treatment with antidepressants may help, to some degree and in ways we don't yet completely understand, to slow microglial damage.

It may well be that the reason antidepressants often take two to three weeks to "kick in" and help some patients is that this is the same amount of time it takes for overactive microglia to back off and allow new neurons to grow, neurogenesis to happen, and the brain to fire and then wire neurons together to make new synaptic connections.*

And yet we also know that the degree to which antidepressants alleviate depression is insufficient to adequately help so many who are in pain to enjoy the relief they long for. More than a third of the 300 million people who suffer from major depressive disorder around the world do not respond to any antidepressant treatments. And some who do respond find that medications stop working for them over time.

It may be that once microglia are engaged in full-throttle runaway inflammation mode, antidepressants, which might help to combat neuroinflammation to some degree, might be too few half-buckets of water thrown on a fire. (And higher doses can increase dangerous side effects.) This may contribute to why so few patients taking antidepressant medications enjoy full, or long-term, remission from the anguish they endure.

---

* In one animal study, researchers found that after only a few weeks of chronic unpredictable stress, microglia can go off kilter and symptoms of depression can appear. When they introduced antidepressants, it helped to block this stress-induced microglial activation.

· · ·

Added to this is the fact that the premise behind SSRIs—that individuals like Katie and Heather have imbalances in specific neurochemicals—doesn't take into account the role that microglia play in brain health. Neurotransmitters, including serotonin, dopamine, acetylcholine, GABA, epinephrine, norepinephrine, and glutamate, bind to receptor sites on neurons, profoundly influencing how well messages are carried along synapses. When levels of these chemicals are altered, over time, brain circuitry no longer functions optimally, and symptoms of depression, dementia, schizophrenia, Parkinson's tremors, anxiety, or OCD can develop.

But there are also problems with this concept that neurotransmitter imbalances are the primary driver of psychiatric disorders.

First, although some individuals with depression do demonstrate low levels of serotonin or other neurotransmitters, many do not. Some have abnormally *high* levels of serotonin. Neurotransmitters may be out of whack, but the resulting chemical imbalances are different in each individual. And that is because neurochemical imbalances are not, in and of themselves, the cause of the disease, but rather a sign of a more profound underlying problem.

The microglial universal theory of disease tells us that when microglia, and the immune system of the brain, sense unmitigated stress, trauma, infection, illness, or toxins, or receive inflammatory signals from a microbial imbalance in the gut, this in turn triggers them to morph from brain savers to synapse slayers, and to release toxic cytokines. These actions alter the availability of neurotransmitters and growth factors, which changes how well signals move between neurons. As the brain's ability to synthesize brain chemicals is diminished, it leads to neural impairments that poison mood, sleep, stamina, concentration, and cognition.

Happy microglia, on the other hand, help to nourish and support neurons and synapses, and this in turn helps to refresh neurotransmitters and maintain them at healthy levels.

Disorders of the mind are first and foremost immune disorders that reflect alterations in the brain's basic immune health.

This is no doubt why the serotonin theory of brain disorders has not really turned out to be the great hope it was initially thought to be: Chemical imbalances are not the root of the problem, but a symptom of it.

We might think of it this way: Microglia are the constant gardeners of our brain's health. They have the ability to turn many different gene functions on or off, depending on the signals that microglia are receiving from the environment. These genes in turn determine the levels of different neurochemicals in the brain. Microglia control these genes a bit the way a gardener controls different faucets and hoses. The gardener can turn up the flow, or turn the spigot back so that the flow diminishes to a mere trickle. How well brain chemicals flow influences whether synapses bloom or wither. But there are so many faucets, and we have no idea how to intervene in a precise way so that they are all turned on to just the right degree in any given individual.

If, instead, we can keep microglia in a healthy state of homeostasis, or reboot them to a healthy state, then we can help ensure that all the faucets are turned on to just the right degree to keep neurochemicals in balance and nurture synaptic health. For this reason, many scientists today propose that mood and cognitive disorders should no longer be viewed as primarily neurotransmitter disorders, but as diseases of microglia and the immune system—or what glial biologists now refer to as disorders of *microgliopathy*.

Not surprisingly, researchers are racing to develop a new class of antidepressants, what we might think of as antidepressants 2.0, that they hope will interrupt the activity of microglia.

Many of these next-generation drugs are specifically designed to either inhibit overactive microglia or stimulate suppressed microglia. For instance, we know that in some neuropsychiatric disorders, microglia become round, oversized, and activated, and they secrete compounds that orchestrate an oversized inflammatory response. But in other cases, too many microglia die off and the remaining cells become small and sickly.

Microglia do a great many different things, at different times, in the face of different environmental stressors—which means that pharmaceuticals that address microgliopathy will certainly not come in a one-size-fits-all package, but will demand a personalized, precision medicine approach and patient-tailored therapy strategies.*

As these explorations into a new class of pharmaceuticals move forward, we will need to proceed with an abundance of caution. Throughout the history of medicine, next-generation pharmaceuticals have often come with a price: serious side effects in some patients. All too often data on these dangerous side effects emerges only decades after the drug has come to market. Clinical trials will need to be repeated and replicated over a long period of time, and in large patient populations, before microglia-targeted medications can be deemed safe for use in clinical practice.

Gene-Targeted Therapies

Since microglia have the ability to turn different gene functions on or off, their actions can dramatically affect any given individual's likelihood of developing a psychiatric or neurodegenerative disorder. So it makes sense that scientists are setting out to target the specific genes that are highly expressed in microglia. Many of the molecular pathways to depression, for instance, including the brain's immune response to substances secreted by gut microbiota, are linked to changes in genes expressed by microglia, which either turn on or turn off neuroprotective or detrimental effects.

In Alzheimer's disease, individuals with a mutation in the TREM2 gene, which prevents microglia from efficiently corralling and clearing amyloid plaques, are, as we've seen, three times more likely to develop Alzheimer's than the rest of the population. But what if we could find a way to safely stop these disease-associated genes from getting turned on in microglia in the first place?

---

* In animal models, when healthy microglia are introduced back into the brain, all signs of depression disappear. Of course, this isn't feasible in humans, but it underscores how profoundly it matters that the microglia population in the brain is healthy.

Let's imagine a teenager who possesses what's known as the C48 gene variant on chromosome 6, the gene known to be most associated with developing schizophrenia and psychosis. This gene predisposes the brain to mark too many synapses with eat-me signals, with the result that too many synapses are destroyed and the brain loses an excessive amount of gray matter. Let's say this teenager's mother has bipolar disorder and her father is emotionally abusive; in other words, her home life is permeated with chronic unpredictable stress. Her genetic predisposition to schizophrenia (since she carries the C48 gene variant) combined with the toxic stress in her home life will vastly increase the likelihood that microglia will switch on any schizophrenia-specific genes she inherited, including C48, as well as stress-specific epigenetic switches. Both of these changes in microglia might set in motion the overpruning of synapses, setting the groundwork for schizophrenia, or other psychosis, to possibly develop.

Investigating how to take a bull's-eye target approach to turn on or suppress microglia-specific genes associated with disease and modulate the epigenome in just such a patient—similar to what is already being done with gene-targeted immunotherapies to treat cancer in the body—is now rich territory for researchers.

Again, we have a long way to go. Researchers are just beginning these explorations and there will be much trial and error, especially given that we do not yet know when it is best to enhance microglial form and function versus suppress it.

Immunotherapies for the Brain

Some labs are working to develop drugs for neuropsychiatric disorders that pivot away from focusing on specific actions of microglia within the brain, and instead flat-out target the body's immune system and thus the underlying neuroinflammation that precedes or accompanies changes in neurotransmitters and synaptic health. The hope is to safely employ anti-inflammatories to target the brain's immune system in much the same way we treat physical inflammation in the body.

The past five years of research in brain-immune science tell us that the body's and brain's immune systems function in tandem. As we've seen, patients with psychiatric disorders often have both higher levels of inflammatory biomarkers in the body and higher levels of abnormal microglial activity in the brain. And if there are ongoing inflammatory triggers—be it emotional trauma or infection—levels of inflammatory cytokines in the brain will continue to provoke microglia to go haywire, leading to neurotransmitter imbalance and runaway neuroinflammation.

Researchers have shown that patients with higher levels of biomarkers for inflammation—including tumor necrosis factor, IL-6, and C-reactive protein—are more likely to be unresponsive to antidepressant treatment. These are very often patients who suffer from the most complex symptoms and conditions: treatment-resistant depression, bipolar disorder, and psychosis. Increasingly, researchers believe that the reason these patients are treatment-resistant, and their symptoms are so debilitating, is because they have higher levels of neuroinflammation.

So it makes good sense that combining anti-inflammatories with antidepressants in patients with psychiatric disorders might one day increase the proportion of individuals who respond to antidepressant treatment and mood stabilizers—or perhaps even circumvent the need for them in some patients.

In 2017, a paper published in the journal *Proceedings of the National Academy of Sciences* showed that when patients with rheumatoid arthritis were given the anti-inflammatory drug Remicade (infliximab), even before clinical measures showed a decrease in their biological disease activity, patients reported positive mood changes—what clinicians commonly call the Remicade high. Taking the anti-inflammatory—which works to block the cytokine known as TNF—also helped their state of mind. Although the anti-TNF medication did not work to treat major depressive disorder in all patients, it did in fact alleviate depression for those patients who also had very high biomarkers for physical inflammation.

At Emory University, Andrew Miller, M.D., director of the behavioral immunology program in the department of psychiatry and behavioral sciences, who, with Charles Raison, M.D., coauthored research examining the evolutionary link between depression, the immune system, and pathogens (which we learned about in chapter 7), has found that patients with high levels of inflammatory biomarkers who are resistant to antidepressant treatment are the very patients who benefit most from infusions of the anti-TNF drug infliximab.*

Another drug used for rheumatoid arthritis, tocilizumab, has been shown to help improve cognition in a small study of patients with schizophrenia. Studies are under way in the United Kingdom to see whether monthly infusions of immune modulating drugs used to treat MS will also prevent microglia from attacking the brain's wiring and improve the lives of individuals with schizophrenia.

But we are still years away from having this kind of precision medicine available at your local psychiatrist's office, and knowing exactly what constitutes ideal versus disease-related levels of inflammatory biomarkers in brain-related disorders, or how to accurately determine just how much of a role physical inflammation is playing in any given individual's psychiatric disease. It will take many more clinical trials before we clearly understand how to test patients' blood in a clinical setting, determine their level of inflammatory cytokines in conjunction with the severity of their depressive symptoms, and potentially prescribe anti-inflammatory treatment accordingly.

Immunotherapies also appear to show possible promise in treating Alzheimer's. Recently, the immunotherapy drug aducanumab helped prompt microglia to turn from destructive mode to happy helpers again, and to start to devour toxins in the brain including the amyloid plaques associated with Alzheimer's disease: 165 patients who received monthly infusions of aducanumab for a year showed a reduction in amyloid plaques, as compared to those who received a placebo. The immunotherapy appeared to slow the progression of the disease in these patients.

---

* Miller is developing more intelligently designed clinical trials, in which patients who demonstrate higher levels of inflammatory biomarkers, including CRP, are selected for inclusion, since patients with depression or anxiety who demonstrate higher inflammatory biomarkers may be more likely to respond to treatment with anti-inflammatories. Miller is also working to bring greater precision to determining which patients with psychiatric disorders will benefit most from dopamine-based anti-depressants versus serotonin-based antidepressants, based on their blood levels of CRP. According to Miller, "Inflammation has an effect on dopamine that leads to many of the symptoms associated with inflammation, including anhedonia. For these patients, drugs that target dopamine may be a better choice of antidepressant therapy."

But the drug was far from perfect: A number of patients dropped out as a result of serious side effects. In some patients, the drug caused fluid to build up in their brain; in others, it led to bleeding in the brain.

As new treatments emerge, we must, again, be cautious. Blocking inflammation can be tricky. Some inflammation can be harmful, while other inflammation is good and necessary; we need our inflammatory response to be up and working to guard against new threats, invaders, and injuries.

Other researchers are examining what happens when antibiotics are added to antidepressant treatment—working with the theory that if the immune system is overactive in body and brain, there may be underlying infections, since inflammation is our immune system's natural response to them and one of the jobs of microglia and the immune cells that signal microglia to become activated is to vigorously fight these off. Small trials utilizing antibiotics such as minocycline, ceftriaxone, and others are under way.

Increasingly, our new understanding of the brain as an immune organ is also informing our insight into so-called nightingale disorders—elusive diseases that rarely get the research attention they deserve because, like the nightingale, who sings at night, the suffering of these patients is so rarely seen or heard. These include disorders such as myalgic encephalomyelitis / chronic fatigue syndrome (ME/CFS), mast cell activation syndromes (MCAS), fibromyalgia, and the like.* We know, for instance, that in fibromyalgia, hyperactive microglia spit out way too much of the cytokine TNF, worsening symptoms. This is also true in chronic pain in general. Among the common denominators in these disorders appear to be neuroinflammation and hypersensitized microglia—so the search for a next generation of pharmaceuticals, which no longer approach the body and mind as separate doors to treatment but instead target neuroinflammation and physical inflammation to address both, may hold hope and promise for these patients.

---

* This term was coined by the filmmaker and ME/CFS health activist Jennifer Brea, who directed the film *Unrest*.

One thing is clear, however: Once we learn how to better help patients, we must intervene early. We know that once microglia go off kilter, without intervention, it can cast a long shadow over an individual's lifetime. For instance, many patients with multiple sclerosis go on to develop dementia or Alzheimer's later in life. Patients with fibromyalgia are twice as likely to develop dementia. Young people who face high levels of anxiety and depression, and who experience some cognitive impairment or learning disabilities, are a staggering 135 percent more likely to develop Alzheimer's as they age.

We are still a long way from understanding how best to modulate the myriad behaviors of microglia—when and how to call them off from harming the brain, when to rev them up to repair the brain, what will work in whom, and why, and when—but many efforts are under way to elucidate answers to these questions.

Can the Biggest Nerve in the Body Reset the Brain's Tiniest Cell?

Scientists are also working to try to reboot the immune system and microglia by hacking the largest nerve in the body, the vagus nerve. The vagus nerve, which is really made up of a bundle of different nerves, is often referred to as the "wandering nerve" because it travels down from the brain stem (just behind that artery on the left side of your neck where you place your fingers in order to feel your pulse) and through your entire torso, sending its roots into your heart, lungs, liver, digestive system, and spleen (a major immune organ).*

The vagus nerve also regulates your autonomic nervous system. Your autonomic nervous system oversees both your sympathetic nervous system, which is responsible for the fight-flight-freeze stress response—your heart rate goes up, you have butterflies in your stomach, blood rushes to your arms and legs so you can fight or flee—and your parasympathetic

---

* The Greek anatomist Galen, a prominent physician during the time of the Roman Empire in the second century, first traced the vagus nerve's path from the brain to the heart and other organs.

nervous system, which regulates how well you're able to rest, relax, and digest, even after a stressful event or perceived threat.

Because the vagus nerve and its branches send nerve impulses and immune signals from the brain into every major organ, the health and vibrancy of your vagus nerve—your "vagal tone"—helps determine how well you respond in the face of inflammatory stressors, including emotional stressors, infections, viruses, and injury—and how well microglia respond too.

The health of your vagal tone is measured (by a physician) by calculating the difference between your heart rate and your breathing. When you breathe in, your heart rate speeds up a bit. When you breathe out, it slows down. The greater the differential between your heart rate when you breathe in and your heart rate when you breathe out, the higher your vagal tone.

In patients with diseases like lupus, rheumatoid arthritis, autoimmune thyroid conditions, fibromyalgia, and chronic fatigue, the vagal nerve often shows a lot of atrophy—vagal tone is very low. Low vagal tone is also strongly correlated with depression, mood disorders, heart conditions, stroke, cognitive impairment, and a range of other physical and emotional health concerns.

Higher vagal tone, conversely, is associated with better mood; stress resilience; less anxiety; lower blood pressure; blood sugar regulation; a reduced risk of stroke, autoimmune disease, and headaches; and many of the factors we associate with better health.

But it was only two decades ago that researchers began to suspect that the vagal nerve played any role at all in the workings of the human immune system. The first glimmer came in the mid-1990s, when neuroscientists at the University of Colorado, Boulder, noticed that when they injected cytokines known to cause fevers into animals but also severed their vagus nerves, the animals did not develop fevers.

At that time, the medical doctrine was that immune cells didn't interact with the nervous system, which includes the vagal nerve. It was a scientific mystery. What did the vagal nerve have to do with fevers, which fell under the domain of the immune system?

In 2000, a neurologist at the Feinstein Institute for Medical Research,

Kevin Tracey, M.D., tried delivering electrical stimulation to the vagus nerve in animal studies, to see if that might also change the way in which the vagus nerve relayed inflammatory signals to and from the brain. He stimulated the rats' exposed vagus nerves with one-second electrical pulses, then injected the rats with a bacterial toxin known to ramp up levels of the cytokine TNF. Normally, injections of TNF would spark full-fledged inflammation and fever. But after vagus nerve stimulation, the rats showed very little inflammation—the production of TNF was muted by three-quarters—which meant that stimulating the vagus nerve prevented the signals that ordinarily ramp up inflammation from getting through.

It seemed the vagus nerve was functioning like a major call center, or phone bank. It could effectively place a call to the brain to say "Hey, we've got a real riot of inflammation going on here, let's get a riot going there too," or it could phone in to say "Everything is fine, no need to get all revved up, relax, and go on about your day."

This relationship between the brain, central nervous system, vagus nerve, and physical immune system is now referred to as the inflammatory reflex: The body and brain sense infection, tissue injury, or inflammation and send this information fanning out throughout the central nervous system, which then reflexively sends a whole host of signals through the vagus nerve and into the organs, telling immune cells to kindle little fires everywhere. The vagus nerve doesn't just communicate with the immune system—it is *part* of the immune system.

Tracey's team recently found that low-level electrical vagal nerve stimulation with implanted devices—zaps of sixty-second bursts, up to four times a day for eighty-four days—damped down the production of TNF and lessened arthritic pain in a very small study of patients with treatment-resistant rheumatoid arthritis. One patient who had barely been able to grasp a pencil prior to the study was, after treatment, able to ride her bike twenty miles.

The vagal nerve stimulator apparatus consists of two parts: a small device that is surgically implanted under the skin along the vagus nerve on the left side of the neck, and a pulse generator that contains a battery and microprocessor. (This is similar to the battery pack that comes with a pacemaker.) Both are connected via a thin platinum alloy wire, similar to the way in which a pacemaker battery box sits near the heart, with lead lines going into the heart to stimulate and correct the electrical rhythm of the heart. (There is one central difference, however: A pacemaker stimulates muscle, not nerves.)

Of course, general treatment with an implantable vagus nerve stimulating device for most diseases in which high levels of inflammation play a role is still far-off-in-the-future medicine—and since it's an implant, it raises many safety concerns. Vagal nerve stimulating devices have been used for some time in treatment-resistant epilepsy, but with a spotty safety record; there have been hundreds of deaths associated with them. And much of this research on vagal nerve stimulation has been done in rodents—which, as we've found out so often in science, are not always a good model when it comes to translating research into humans. Few human trials have been done, and those that have been done were in tiny groups of patients who have not responded to other previous treatment approaches—in other words, patients with nothing else left to try. And, according to researchers at the Georgia Institute of Technology, in some instances, vagal nerve stimulation may do the opposite of what it's intended to do and escalate inflammation—potentially harming some patients. So they are experimenting with delivering low-level vagal nerve stimulation while simultaneously injecting a nerve block, to see if they can modulate the potential negative effects of vagal nerve stimulation.

Still, since microglial immune cells are affected by electrical messages received by the brain, it makes sense to try hacking the biggest nerve in the body to see if it has an appreciable impact on the microglia-body feedback loop that drives inflammation in diseases of both body and brain. A whole host of studies are now examining, in animal models and some small human clinical trials, whether vagal nerve stimulation can help ameliorate symptoms and inflammation in colitis, Crohn's disease, sepsis, diabetes, chronic pain, pelvic pain, fibromyalgia, headaches, epilepsy, heart failure, memory loss, and depression.

Researchers in Denmark are also looking at how unhealthy changes in the gut microbiome signal the brain through the vagus nerve, which, they believe, may be a precursor to Parkinson's disease. The Harvard researcher Michael VanElzakker, Ph.D., is investigating whether it's possible that whenever the immune system senses any peripheral infection, the vagus nerve detects inflammatory signals, and, acting as an immune conduit to the brain, sends a signal to microglia to initiate the symptoms of fatigue, depression, and sickness behavior—symptoms that overlap with chronic fatigue syndrome. For instance, if the vagus nerve detects the presence of an unhealthy microbial population in the gut, it passes that information on to microglia, stirring up neuroinflammation. VanElzakker is also examining whether low-grade infectious pathogens and viruses, including those of Epstein-Barr and Lyme disease, might hide within the bundle of bidirectional cranial nerves that make up the vagus nerve, continually triggering microglia and the peripheral immune system to go haywire and unleash symptoms of malaise. He believes that when the vagal nerve itself is infected with a virus or bacteria, this causes the body's immune cells to trigger a mirror response in the brain, activating microglia to bombard the brain with more inflammatory cytokines, in what VanElzakker calls "an exaggerated and intractable sickness behavior signal."

It is as if once this vagal nerve cross talk arises, the body and brain throw down a red flag at the starting line and inflammation revs up its engine. The question is whether stimulating the vagus nerve can safely slow that engine down.

Hallucinogenics

Yes, you read that right. The hallucinogenic ketamine has long been used as an anesthetic in hospitals, burn centers, and veterinary practices. Increasingly, ketamine also appears to have the potential, in very low doses, to confer a strong antidepressant effect because of the way in which it inhibits the ability of microglia to release toxic, inflammatory cytokines in the brain, thereby allowing neurogenesis, and new synapse growth. Although these microglia-specific findings have been in animal studies, the effectiveness of treatment with low-dose ketamine, delivered intravenously, is backed up by recent human clinical trials.

Ketamine appears to confer its anti-inflammatory and antidepressant effects by modifying microglial behavior in a way that leads to changes in levels of the neurotransmitter glutamate, which, in turn, helps to restore and turn on synapses and circuits.

Susannah Tye, Ph.D., former director of the translational neuroscience laboratory at the Mayo Clinic, who now serves as group leader of the functional neuromodulation and novel therapeutics laboratory at the Queensland Brain Institute in Australia, recently co-founded a task force among hospitals in the United States to focus on improving treatments for depression. This consortium, known as the Treatment Resistant Depression Task Group—part of the National Network of Depression Centers—recently launched a multicenter biomarker trial, known as the Bio-K Study, to look at the efficacy of ketamine in treatment-resistant depression and bipolar disorder. Hospitals in the consortium, which include Mayo Clinic, the University of Michigan, and Johns Hopkins, are investigating whether patients with higher blood-based biomarkers for inflammation might benefit most from ketamine treatment. Patients' biomarkers for inflammation are tested before they receive three low-dose IV ketamine infusions. After each treatment, patients are evaluated for changes in depressive symptoms, motivation, and energy, along with changes in blood-based biomarkers. By utilizing patients' biomarkers to help determine if ketamine treatment is effective for them, this work, says Tye, "is leading the way toward new, individualized medicine approaches for psychiatric disorders." As research and evidence are gathered, it will become more likely, she believes, "that patients who have high levels of inflammation will be directed to treatments that can directly target this pathophysiology—treatments such as ketamine or other emerging anti-inflammatory agents." Thus far, these studies are small but promising.

The hope, says Tye, is that "the work investigators are doing to apply our understanding of the role of microglia and inflammation in treating psychiatric disorders will also translate into the field of psychiatry and clinical practice." As it stands, she adds, "Your typical doctor or psychiatrist may not be aware that general inflammation may affect treatment outcome. A psychiatrist may offer a diagnosis of major depressive disorder in three patients, all of whom may have very different physical profiles. If we target treatment in the same way in all of them we won't be targeting the right pathology." We are, she believes, "still going more by the label of the disorder than by the biology of the patient.

"Our understanding of microglia and neuroinflammation in basic brain function and in psychiatric disorders improves year by year—it's the new frontier of medicine," Tye adds. "We have the knowledge base now to begin to change the way we are treating patients. We are closer than we have ever been. But this will require that psychiatry embraces the rapidly developing knowledge coming out of the field of neuroscience and the role of this little cell in healthy brain function and in psychiatric disorders."

Some physicians aren't waiting for all the clinical trials to be completed and are now offering low-dose ketamine IV treatments to patients in their clinics, claiming as high as a 75 percent success rate in patients for whom antidepressants, mood stabilizers, transcranial stimulation, cognitive behavioral therapies, and other treatments have failed.

The former director of the National Institute of Mental Health, Thomas Insel, has gone so far as to state that "recent data suggest that ketamine, given intravenously, might be the most important breakthrough in antidepressant treatment in decades." When ketamine works, it doesn't take weeks to offer relief, either; patients report that they experience a lightning-swift change in how they see the world within minutes or hours.

But there are major downsides. Ketamine (in much higher doses) is also known as the party drug Special K, and as a so-called date rape drug because it can quickly render a person unable to move. It is therefore a highly controlled substance, which can be safely administered only in a doctor's office. And although the very low doses at which infusions are delivered to patients with severe depression have not led to the addiction or cognitive decline that's been reported among those who abuse higher doses of ketamine as a party drug, in some patients even low-dose treatments have caused them to dissociate, experience lucid dreaming, and hallucinate.* This is not surprising, given that about half of patients who receive high doses of ketamine as an anesthesia in a hospital setting experience the kinds of "emergence phenomena"—symptoms that emerge as a patient comes off of anesthesia and returns to consciousness—that can mimic psychosis.†

Ketamine treatment is also prohibitively expensive. Currently, it requires three IV drips a week while being supervised in a doctor's office, which comes at the price tag of roughly $500 per infusion—often more. That cost can diminish with treatment, with some patients reporting that over time the drug's effects last for much longer periods, so that they are able to eventually go weeks or even months between infusions.

It's also hard to find doctors who administer ketamine. In part, that's because the physicians who are most familiar with administering ketamine are anesthesiologists, who are unlikely to treat psychiatric patients. Meanwhile, psychiatrists, who are well schooled in prescribing pharmaceuticals to address possible neurotransmitter imbalances that they hypothesize may be occurring in any given patient's brain, are unlikely to be hooking patients up to IV treatments anytime soon.

Meanwhile, many patients who have sought off-label treatment for depression with low-dose ketamine infusions have reported recovery from even intractable cases of depression, bipolar disorder, anxiety, PTSD, and OCD. Many say that the drug not only rapidly relieves their mental angst but helps them feel a greater sense of connection to other

---

* Patients with depression or bipolar depression with a history of psychosis may experience a greater sense of dissociation during the forty-minute infusion treatment. However, recent small studies show that after treatment, their symptoms improve.

† Interestingly, patients who experience "emergence phenomena" after being given ketamine as an anesthetic in a hospital setting may be more likely to possess genotypes in receptors for the neurotransmitter glutamate that are also associated with ADHD, epilepsy, and schizophrenia (which is not to say that they suffer from these disorders).

people and to the world around them. (Perhaps part of the drug's psyche-delic "we are all one in the universe" effect.)*

Another hallucinogen, psilocybin, also known as magic mushrooms, is being studied in clinical trials and appears to help ameliorate symp-toms of depression and anxiety in patients with late-stage cancer. We don't yet know why or how it works, but psilocybin seems to disrupt the negative, self-focused ruminating that is a hallmark of depression and anxiety, helping patients to experience a more mystical sense of the world in a way that relieves mental duress during cancer treatments.

Other investigators are looking at a beverage used for centuries by na-tive South Americans, ayahuasca, made from an herb that has also been shown to help promote neurogenesis in the brain, which also suggests that toxic microglia are backing off. Studies are under way to evaluate whether ayahuasca can help patients with depression too.

A Potent Combination Effect

Just as disease in any given individual is multifactorial, so is recovery. As with most stories of successful healing, patient recovery will most likely come when individuals take a combination approach, utilizing a num-ber of current and emergent modalities to achieve better neuroimmune health. Patient "A" might find relief, one day, in combining ketamine and a fasting-mimicking diet. Patient "B" might benefit most from com-bining immunotherapy and transcranial magnetic stimulation. These might be done in conjunction with antidepressants, while monitoring levels of neuroinflammation.

---

* In 2019, the FDA approved the use of a nasal spray made from a ketamine deriva-tive, esketamine, for patients with treatment-resistant depression. But there are con-cerns: The first month of treatment alone may be more expensive for patients than IV ketamine treatments. Like IV infusions of ketamine, the nasal spray may cause out-of-body sensations and hallucinations, and has the potential for abuse. For these reasons, it, too, must be administered in a doctor's office and patients need to be observed for several hours afterward. And, because the drug is brand new, we do not yet know its long-term efficacy or potential risks.

Once all of these approaches are rolled out in a way that is safe, widely available, covered by insurance, and affordable, we can begin to see what is working best in which population, at which ages, for which genotypes, for which disorders. Perhaps scientists can then go one step further and study what combinations of approaches create the most powerful healing effects of all.

This is science in motion, but the good news is that the field is moving incredibly fast. We are about to enter a new era, one in which we will not replace long-standing and crucially important therapeutic approaches such as psychotherapy, CBT, mindfulness meditation, trauma-aware interventions, current generation antidepressants, and other interventions that encourage neuroplastic change, but will add in new approaches that draw upon the power of microglia to recalibrate the underlying health of any given individual's brain throughout the life span. Approaching the brain as a unique immune organ doesn't negate current approaches to alleviate human suffering; rather, it offers us insight into a new set of potential tools in the toolbox, at a time when there is such an outcry for relief.

SIXTEEN

A Final Analysis

SINCE THE ENLIGHTENMENT, MUCH OF WHAT SCIENTISTS, CLINI-cians, and patients have assumed to be true about the intersection between the brain, the body's immune system, and human suffering has been based upon the age-old assumption that the workings of the mind are divorced from the mechanisms of the body. The philosopher Descartes first put forth this concept, known as mind-body dualism, in the seventeenth century, and the resulting medical dogma that mind and body function as separate entities has permeated the way in which we view and approach disorders of the human mind long into the twenty-first century.

We've become so deeply habituated to thinking about the workings of our minds and bodies in a church-and-state way that when something goes amiss in how well we are able to manage our mood, in how calmly or explosively we find ourselves reacting to the world around us, or in how vibrantly we are, or are not, able to engage in this life, we think primarily of our mental and emotional state of being. We may, like Katie, Heather, and Lila, begin to judge ourselves harshly for our entrenched depression or anxious state, our forgetfulness, our dearth of joy, inter-weaving our disappointment in our state of mind into a systemic disap-pointment in *who we are*. This in turn can rewrite our sense of self until we feel we are someone very different from who we might once have been, and a thousand miles from who we set out to be.

But our new understanding that the brain is a sensitive immune organ, constantly on the lookout for possible new threats—and that myriad immune triggers can slowly change the habits of microglial cells in the brain, so that they remodel our synapses in suboptimal ways, just as environmental triggers can alter the habits of immune cells in the body—erases this three-hundred-year chasm in our fundamental understanding of mental health.

As Jony Kipnis at the University of Virginia recently put it, "We have five established senses—smell, touch, taste, sight, and hearing." Our sense of proprioception, or where our body is in space, is often thought of as our sixth sense. These six senses "report to the brain about our external and internal environments, providing a basis on which the brain can compute the activity needed for self-preservation." Kipnis proposes that one of the defining roles of the immune system is to detect when there are threats in our environment and to constantly inform the brain about them, and if, as he theorizes, this "immune response is hardwired into the brain," that would make our immune system's constant back-and-forth communication with our brain our "seventh sense."

Researchers at the University of Colorado, Boulder, recently furthered this understanding when they demonstrated that the central nervous system can respond to even emotional stress "as if it were cellular damage," releasing danger signals in the brain that prime microglia to behave as they would if they were fighting off an infectious immune challenge.

This concept—that the brain's immune system functions as our seventh sense when it is under a perceived threat and may begin to orchestrate a cascade of changes that slowly set in motion a range of disorders of the mind—offers patients a new understanding of their conditions while simultaneously offering physicians and practitioners an entirely new playbook on how to intervene and help patients to reclaim well-being.

This astonishing research of the past decade, what we might dub the *real* Decade of the Brain, changes the basic ideas that have been instrumental in how we view who we are and why we are the way we are.* This is really nothing less than a new window into the making and unmaking of the self.

There is, of course, a risk in focusing our attention on the brain as an immune organ. If we overemphasize the workings of microglia, and the biological mechanisms by which illnesses of the brain emerge, we invite the kind of biological reductionism that overmedicalizes and belittles the intimate connection between the mind and the way it gives birth to our human consciousness. But this science is not meant to try to decon-struct the unfathomable complexity of each human being's inner life, or the nature of a healthy mind, which can never be reflected in PET scans, or by blood biomarkers, or by unveiling the actions of one cell. One's emotional or spiritual experiences as a sentient human being can never be explained by a set of purely cellular processes. The power of human connection and relationships in shaping our minds and souls, addressing our emotions and trauma, and unraveling our interior feelings of grief and loss is crucial to every patient's healing. Increasingly, we are also coming to understand that the caring and empathic quality of the patient-healer relationship greatly increases the biological healing effect and outcome of medical treatments. This effect can never be reduced to simply mechanistic biological explanations.

As the pediatrician Robert C. Whitaker, M.D., M.P.H., director of re-search and research education at the Columbia-Bassett program, wrote: "We can understand a lot about the immune system, it seems, by study-ing it 'outside' the body. We cannot understand the mind outside the body. In fact, the mind is not just the brain and we cannot remove the brain from the body to even understand the brain." We are, says Whita-ker, "still trying to understand the nature of a healthy mind."

Similarly, this science should not be used in a way that leads us to catastrophize our brain's fragility by assuming the mind is primed to dis-mantle itself and thus underestimating our brain's innate potential for

---

* Former U.S. president George H. W. Bush dubbed 1990–1999 "the Decade of the Brain," in conjunction with the National Institute of Mental Health of the National Institutes of Health, in order "to enhance public awareness of the benefits to be derived from brain research."

healing. Rather, understanding the promise and peril of this science tells us that our emotional and physical capacities cannot be separated: The emotional gradually becomes physical, and the physical gradually becomes emotional.

This tiny cell, the power of which science overlooked for so long, plays some role in every story of human suffering—or in what we might think of as the disappearing self.

Nearly a century ago, Sigmund Freud, the father of psychoanalysis, cautioned against bringing medicine and psychology together, warning that psychoanalysts needed to glean all they could from "the mental sciences, from psychology, the history of civilization and sociology" and not from "anatomy, biology and the study of evolution."

But a hundred years later, these two roads are converging, requiring us to draw urgently upon all that science knows in order to fully bridge the three-hundred-year gap between radically different understandings of mind and body. If any given patient's neuroimmune system begins to shift in ways that alter her or his possibilities and passions, we can now hope one day to use new techniques to help shepherd microglia back to guardian-like roles. We are learning how to restore their natural capacity to nurture the brain throughout the many seasons of an individual's life, early development, adolescence, midlife, and old age.

Descartes—and three centuries of medical dogma—could not have had it more wrong.*

Clearly, our narrow medical approaches toward treating disorders of the mind have to date been only marginally successful. Many individuals

---

* A few interesting sidenotes here: Although Descartes gave us the dualistic view of body and mind, ancient Greek philosophers were fascinated with distinctions between psyche, mind, body, and other parts of the self. While some believed that thought processes were located in the brain, Aristotle speculated it might be a cooling organ. Galen later promoted the brain as ruling a complicated system integrating sensation, thought, and emotion. Also, Descartes's severe separation of mind and body (which included locating the soul in the pineal gland!) was challenged by a woman, Princess Elisabeth of Bohemia, with whom he engaged in a lively correspondence.

who experience depression, mood disorders, and cognitive decline feel that current approaches to improve mental well-being are not enough, in and of themselves, to alleviate their distress.

As patients like Katie can all too well attest, although antidepressants and mood stabilizers can be lifesaving, especially during acute psychiatric freefalls, they bring with them a whole host of side effects that make life difficult to navigate in new ways. For so many, these therapies are far from enough to help them feel that they are in love with this life, or to savor the joy of living.

Even many psychiatrists will admit that the field of psychiatry has fallen starkly behind that of other medical disciplines. Survival and recovery rates for other serious medical conditions, including heart disease and cancer, have improved greatly in the past fifty years. Meanwhile, recovery rates for mental illness have barely budged, and we have made almost no progress in halting neurodegenerative disorders such as Alzheimer's, even as the numbers of those afflicted with these disorders continue to rise.

Perhaps nothing speaks more starkly to the insufficiency of current treatments than the terrifying fact that over the seventeen years between 1999 and 2016, suicide rates have risen in the United States—making it the second leading cause of death in those ages fifteen to thirty-four and the third leading cause of death in children between the ages of ten and fourteen. And among those who commit suicide, nearly a quarter are already taking antidepressants.

It is my hope that this new understanding of the role that microglia play in disorders of the brain will not only help patients to be aware of new treatment options, but also help destigmatize their suffering. The microglial universal theory of disease helps us to recognize that we are *all* dependent on the health of our brain's immune system, just as we are *all* dependent upon the immune health of our bodies.

Beth Stevens sums it up thus: "We have been throwing drugs at the brain without entirely knowing what those drugs are affecting." And this has added to patient frustration and confusion. But, Stevens offers, "As we further pinpoint the genetic mechanisms and pathways that indicate that something is going wrong in the brain, and create a roadmap of the genetic interactions that are so often at the heart of these challenging neuropsychiatric and cognitive disorders, this will also help to destigmatize these diseases, by helping patients to understand that these genetic interactions are *not* their fault, and these disorders are not their fault. We have never before had this kind of concrete understanding."*

Biology is complex, and the brain's health is contingent on so many different emotional and environmental stressors, coupled with genetic predispositions. Newly categorizing psychiatric and neurodegenerative disorders as also being disorders of microgliopathy and the immune system is useful for furthering research and understanding.

And that's because the language we have long used to describe disorders of the brain has become dangerously outdated. As such, it can interfere with the decisions that patients and their physicians make regarding patient treatment options. Understanding that psychiatric and neurodegenerative disorders are also disorders of microgliopathy and the immune system matters in terms of catalyzing physicians and psychiatrists toward a new awareness as they work with patients who are desperate and hopeful for new answers.

The language we use also makes a difference in terms of funding—which is desperately needed. At one 2017 gathering of the top glial biologists and neuroimmunologists from around the world, scientists lamented the "limited funding currently available for research on neuroinflammatory diseases."

Meanwhile, disorders of the mind and brain bring with them a heavy cost burden. In 2013 (the latest year for which we have numbers), the United States spent $201 billion on treating mental illness—more than on any other disease, including cancer, heart conditions, and diabetes. Most of this burden falls on families—precisely because we have decided that these disorders are not physical and therefore don't need to be fully covered by "medical" care. Far too few psychiatrists, therapists, or

---

* The use of the term "genetic" here refers to both inherited genetics and epigenetics, or the ways in which our genes change in response to our environment.

psychiatric treatment centers accept patients' insurance, leaving patients and their families tired and tapped out, emotionally and financially.

Yet even as American families spent more on mental illness than on other diseases, the United States spent fewer research dollars on figuring out how to improve treatment for mental illnesses, as compared to cancer and heart disease.

If we are to help patients, we must fund the researchers who are trying to design and achieve safe new treatments so that patients can thrive.

And we must demand that our health care system utilize this information in ways that give patients who are suffering more practical knowledge, options, and power.

We stand at the cusp of a sea change in psychiatry: an enormous paradigm shift that cuts across all areas of medicine, and promises to rewrite psychiatry as we know it—based on the novel understanding that microglial cells sculpt our brain in ways that have profound lifelong effects on our mental health and well-being.

Hope for the Future

Microglia are, in simplest terms, both the assassin of the self and the guardian of the self. And science may be able to help us bring them back into homeostasis so that those suffering with depression, anxiety, obsession, distraction, or forgetting can finally escape the thieves that can rob them of whole lifetimes. It has been said of individuals with psychiatric disorders that time has "stopped inside their wounds, which are seemingly never to heal."

Researchers like Beth Stevens and her colleagues are looking ahead to the next ten years, and to where this science will take us, in hopes of offering individuals and their families new visions of hope. In order to achieve this, Stevens says, "The most important thing is that instead of scientists staying in their different silos, we all need to come together and share our data, protocols, and studies with complete transparency, even before we wait for our data to be published, so that we can move forward collaboratively toward a goal beyond ourselves with greater speed. Team science is what it's going to take to achieve this big mission, if we are to make an impact on people's lives, and on society."

As for Stevens herself, "We're just getting started," she says. "We have so much more to do. We can only imagine where the science will take us from here. Ten years ago, I could not have imagined, when we started looking closely at microglia, that studying microglia in normal brain development might be relevant in unraveling the mystery of psychiatric and Alzheimer's disease. I never imagined that that research would lead me here." In the next five years, she hopes, researchers will be able to provide many more answers for patients that will surprise us all.

This will require disruptive innovation and open collaboration: a convergence of the fields of neuroscience, genetics, psychology, psychiatry, medicine, and immunology—a recognition that they are all one field— tied together by one tiny cell that has altered the course of our understanding of the human brain, in ways that will translate into helping us all to live our most capable, meaningful, satisfying, and transcendent human lives.

Epilogue

OW HAS THIS NEW UNDERSTANDING OF BRAIN HEALTH SERVED Katie, Heather, and Lila? In evaluating any treatment that's thought to help influence the behavior of microglia and target the brain as an immune organ, it's important to investigate not only whether an approach helps in the short term, but also whether progress continues to hold over the long term.

Katie Harrison

Katie Harrison and I meet one last time, not far from Dr. Hasan Asif's Bronxville, New York, office, where Katie, who is back in town to see Dr. Asif for the first time in six months, has just received several transcranial magnetic stimulation treatments as a "tune-up." It has been a year since we first met. We sit on a bench at the Brooklyn Botanic Garden, beside large rectangular pools blooming with water lilies and lotuses. A few ducks wander in the bordering garden. It's a crisp but warm fall morning.

The first thing I notice is that Katie has recently had her hair cut shorter, and she has color in her cheeks.

"You look so different!" I say.

"I was wandering around Bronxville yesterday, and I went by a salon," she says. "I popped in and asked if they had any openings. I got my hair cut and blow-dried!" Behind her, ducks waddle and peck their way through the garden. "I used to cut my hair myself, because when I went into a salon, I felt as if the inside of me and the outside of me didn't match. My hair would look better, but as I looked in the mirror I'd be thinking about how incongruous it was, because inside I still felt so much self-loathing."

It's then that I realize Katie is, perhaps for the first time since we met, looking me in the eye as we talk.

I'm not the only one who's noticed significant changes in Katie's demeanor. She tells me she recently ran into a friend at the grocery store who told her, "Katie, you look so . . . alive!"

Katie's days, she continues, are entirely different. "The other day a friend called and asked if I could watch her kids for an hour," she explains. "When I said yes, my friend hesitated and said, 'Katie, really, are you *sure*? I hesitate to take you up on this because I know in the past how hard things have been for you.' I told her, 'I know, I know, but I'm feeling better. Besides, it will be fun!'" She laughs a delightful, musical laugh. "When they all came over, I put all four kids in the car and we went out for ice cream. When my friend came to pick them up, we were outside playing cops and robbers. My friend said, 'My gosh, you've totally changed. You must *really* be feeling better.'"

Not long after, Katie decided to travel with another friend, who is also a single mom, and their four kids, to Legoland. "The last time I was at an amusement park, four or five years ago, I was not able to enjoy it at all," she recalls. "I was constantly worrying that I was going to lose one of the kids. This time, the thought crossed my mind, but then I let the thought go. I didn't worry about it again."

When Katie and her kids got home, her part-time babysitter broke it to her that she was moving out of the area. "I realized that if I could take the kids to Legoland, I didn't need a regular babysitter anymore." Katie grins. She starts to list the things she's able to do now, ticking them off one by one on her fingers: "I'm getting the kids organized for school every morning, running errands, having coffee and lunch with friends, setting up playdates, going to therapy, watching my son's karate competitions, rooting from the stands. I used to have to sit in the car during them; I couldn't handle the din and noise of the gymnasium. I hadn't been able to go to the dentist for years, because it made me too anxious, but I went a few weeks ago, and I was fine! Next Saturday, a friend and I are taking our kids to the water park. I'm doing more by myself than my babysitter and I used to do together!"

In the past, Katie admits, "I knew my kids were playing too many video games. But I didn't have the mental energy to do anything about that. It was the best I could do. I had to almost encourage it because I *needed* the quiet." But now, she says, "I have the energy and mental bandwith to be the parent I *want* to be."

Sometimes, she says, out of habit, she hesitates to do something, and then pauses and thinks, "Really, it's fine, I can do it!" For instance, "I've always wanted to go to estate sales, but I would feel too anxious to go in. Last weekend, while the kids were with their dad, I drove past a house with an estate sale sign in front of it and lots of people were coming in and out. I went in and shopped around. I felt comfortable. I even had fun chatting with people."

Katie's daughter, Mindy, had begged Katie to be able to join Girl Scouts for years. But the meetings were at night "and that was out of the question for me," Katie says. This fall, Katie took Mindy to their first meeting. A few weeks later, when the Girl Scout leader announced she was going to be on bed rest for the duration of her pregnancy, Katie volunteered to become troop leader. "Now I'm a Girl Scout leader!" she says. "I used to assume that I just had no zest for life, but that is not the case at all. It was just buried beneath my depression and my sense of panic. To be able to do these things now is *exhilarating* for me."

She is quiet for a moment as we simply sit, feeling the slight breeze play across our faces. The water lilies and lotuses have opened to the slanting morning sun. "I used to have this weird thought that would permeate my mind all the time," Katie confides, her voice lowered. "I'd see moms and their kids having fun, or women out to lunch together laughing, and I'd think to myself, *Real people do those things.* I felt that somehow *I* wasn't a real person. The best way I can sum it all up is to say that before I was pretending, as best I could, and now I finally feel like I'm a real person."

Katie's newfound mental and physical stamina isn't life-changing just in terms of trying new things. "It's also about enjoying life's quieter moments," she says. "Mindy and I have been gardening this past summer. We grew tomatoes and cucumbers. We had so much fun harvesting them together and making gazpacho. It was just a very joyful thing to do with my daughter."

Recently, Katie tells me, she put photos into all of the empty picture frames that have been sitting on tabletops and in bookshelves around her house for years.

"You've had empty picture frames sitting around without photos in them?" I ask.

"Yes," she confesses. "I just never put photos in them. What was I going to fill them with? But now, I look around and see these photos of a person—and a family—that is living a completely different life from the family we used to be.

"I'm feeling really good physically too," she says. "I used to slow jog or walk every day for twenty minutes and I'd just grit through it. Sometimes walking aggravated my fatigue and made me feel so much worse. Especially if it was hot. I'd feel like I was going to collapse. Now I'm going to the gym twice a week. I'm *physically* stronger."

Looking back, Katie admits she sometimes feels "grief for all the time I lost that I can't ever get back. All the missed opportunities as a parent, in my work, my relationships. I can never get it back again." The other grief she feels is for the way she judged herself constantly. "I didn't understand why I couldn't do what other people could do, so I'd lambaste myself. I'd look at everyone else and think, well, no one *else* has a problem being a parent, or having friends. It must be just me. I felt like a waste of space."

Katie has even set steps in place to go back to work. "One of my friends has a psychology practice in northern Virginia, and she asked me to give a presentation about what I know about TMS for her employees. I had to read a bunch of technical books and put together all the science. I gave a PowerPoint, and told my story, and they loved it!"

"You gave a professional talk in front of a group of professionals?"

"Yes!" she says.

"Do you want to go back to work as a licensed social worker?"

"I do," she says. "I've thought a lot about what it means to be sick in our society, and how hard it is to feel good about yourself when you suffer from a mental health disorder. When you are deeply depressed, you start to feel like a waste of intelligence, of ability, of possibility, that you are not enough and never will be enough, that you will never be yourself again and that no one can ever fully grasp the sense of hopelessness you feel. And all these thoughts are overlaid with a kind of fatal exhaustion. Having a new sense of possibility, a new way of looking at depression, *that* is a light in the dark. I want to get back into practice and help shed that light of new possibility in other people's lives."

"Are you taking care of yourself in the midst of doing all this?" I ask.

"Yes!" she says. "In the past I stayed on a strict diet, and exercised, and rested because I *had* to. It was forced self-care. It was a job I had to do, just so I could function at some minimal level. But now I'm starting to take care of myself because I want to. It doesn't feel like work anymore." For instance, she explains, "This past weekend my kids were at their dad's and I cooked my favorite foods and read fantasy books all day long on Saturday. I haven't done that since I was a child." A smile breaks over her face.

"What did you read?" I ask.

"The entire *Wrinkle in Time* series!" She laughs.

"How often will you still see Dr. Asif for follow-up treatment?" I ask, curious about how much maintenance TMS therapy Katie will need.

"Every six months." She smiles as she adds that Dr. Asif estimates that in another six months' time, Katie will need to see him only once a year or so.

"What about medication?" I ask. The last time we spoke, Katie told me that she had been able to cut the dosages of the antidepressants and mood stabilizers she'd been on for years by half.

"After twenty-five years I'm off *all* of my medications," she says. "Occasionally, I have to take something to help me sleep, but that's now the exception rather than the norm. I used to travel around with a small pharmacy."

All of these changes, Katie says, have altered not only how she sees herself, but the way in which she sees others. For instance, she explains, "I used to have this idea that my parents didn't understand me, or grasp how much I was suffering. And to some extent that may have been true. But now I can see how much they have always tried to help me, to make a difference in any way they could, in the best ways they knew how. Now that I'm better, they have been able to share the agony that they felt in watching me go through so much suffering. We are so much more honest with each other now. We're a lot closer."

Katie believes it has also helped her family "to better understand exactly what has been happening in my brain; that these little immune cells have been overactive in my brain, in the same way that their immune cells may have been overactive in different organs in their bodies." With this new understanding of the brain as an immune organ, and the role that microglia play, she says, "They really get it. I get it. It's such a difference in perspective. This science eradicates so much blame and shame."

It occurs to me, as Katie talks, that she no longer seems to ache with a sense of dislocation.

"Illness brings everyone down," Katie muses. "And when the person who is hurting gets better, it brings the whole family up again. My parents, my kids, they are all lighter and happier. Hasan Asif didn't just save one person, he saved five of us."

Heather Somers

Heather Somers and I enjoy our last catch-up in a Baltimore bagel shop. It's been nine months since she completed her twenty-four qEEG neurofeedback sessions with Mark Trullinger. As she sits down, she starts to tell me about the most exciting change that's occurred in her life since undergoing treatment. She has been able to get her school wellness program, which is focused on girls' self-esteem and social media, off the ground. And it's grown into something that has "real legs," she tells me.

She describes her program, Big Sisters on Social Media. The initiative, which pairs high school senior girls with middle school girls, asks seniors to act as the younger students' social media big sisters. "They meet at the beginning of the school term, during wellness class, and create short social media film clips that touch on advice about how to use social media, what lessons the older girls have learned about the benefits of limiting their time on it, and any regrets that they might have, that they feel the younger girls might learn from. All the clips go up on our group's private web page." Throughout the year, "if incidents of bullying or peer pressure or body shaming come up, or if there is something the younger girls don't know how to handle, big sisters step in. If the older girls need help handling a situation, we bring in the school counselors." Heather has just come back from giving a talk at another school about her program's success, and she is brimming with excitement.

"Being able to achieve this really reflects two huge shifts in how I feel," she says, taking a hearty bite of a gluten-free bagel sandwich piled high with lox and cream cheese. "The main thing is that I feel mentally *organized* and focused enough to have turned my ideas into real programming, and pilot it. I can't imagine having done this even nine months ago—I didn't have the attention span. And now I'm thinking about how to scale it all up!"

The second interior shift, Heather says, is harder to define. "I feel as if for the first time in my life it's okay for me to focus on being happy, even if and when other people in my life aren't feeling well. That's new for me. I've always attached my happiness to whether the people I loved were happy and doing well."

After her recent work trip, Heather tacked on a few days at the end, to do something she'd always wanted to do. "I realized that the school where I was lecturing was half an hour from Fallingwater, the house that Frank Lloyd Wright designed. I've always wanted to see it. I hesitated to allocate the time and money to travel for myself, but then I just did it. I booked the hotel for an extra night and went to see it. It was great!"

I ask her about the issue that had been, I knew, the most difficult for her in the past year. "How is Jane?"

"She is in school, doing better—she's meeting with a new neurofeedback expert near campus," Heather says. "It's helping, little by little. Oh, and Jane worked as an intern for Big Sisters on Social Media all summer! She really got into it." She gives a small smile.

This past week, she says, "Everyone was talking about plans for Christmas at our house—my parents and Dave's parents—and I was feeling overwhelmed. I have always hosted Christmas. I'll spend weeks preparing and shopping and cleaning and cooking so that every minute is great for everyone." Heather was able to be utterly honest with herself and recognize that this year, she didn't want to "be caregiver and housekeeper for seven grown-ups," she says. "It was really fun when the kids were little. But now it's so much work and I have so much else on my plate. I just want to enjoy the holidays, and being together, without waiting on everyone." So Heather went online and booked a resort in New Hampshire; she and Dave, her parents, and her in-laws are all sharing the cost.

In the past, Heather confides, "those feelings would have been so conflicting and sticky for me. The biggest change for me is that my passive-aggressive feelings just melt away. I notice I am ramped up, and I see that this is an opportunity to figure out what it is that I need. And once I do that, I'm able to focus more on deciding what I want to do while also being more loving and present with the people I care so deeply for. And they are able to be more loving with me."

Lila Chen

Lila has gone on the fasting-mimicking diet twice more since she first tried it, per Valter Longo's protocol, which suggests that the ProLon FMD be done once a month for three months.

Each time Lila has done the FMD, she's found it to be a little more beneficial. "The first time I felt as if the energy bump and added clarity lasted about a week, then it began to fade a little," she tells me as we take a weekend walk in a local park. "It was almost like a temporary high. But each time I've done it since, I've felt better for a little longer. Because I'm doing a little better overall each time I start it, the benefits seem less extreme, but they also seem to have more cumulative staying power."

Lila says that the new clarity of mind she's experiencing feels like a very different way of being in the world. "I'd sometimes think, wow, here I am, someone who aced the GREs thirty-some years ago, and now all of that intellectual ability, all the strengths I worked so hard to develop in college and grad school and early in my career, those skills are just *gone*. When I couldn't figure something out for myself, like how to work the GPS in my car, or how to download a podcast, or if I had to read a paragraph eight times, I just accepted that I've lost my intellectual edge, that acumen.

"I thought I'd realized how diminished my cognitive abilities had become, but it wasn't until things became clearer in my head that I realized how bad things were. I was in such a daze. Such a fog. It was like I was piloting a plane with a damaged radar screen, and I could only see a small radius around me—just enough so that I didn't crash and just enough to know where I'd recently been. I had a hard time accessing all of the information I needed to plan ahead and make decisions." Each time Lila has done the fasting-mimicking diet, she says, "It feels as if I've cleaned up a little bit more of that radar screen."

Lila smiles and waves to a few women we know who are walking past us in the other direction. She calls out their names as she says hello.

"See that?" she asks, glancing at me. "I haven't seen them in four or five years, but their names came right back to me. That would never have happened six months ago. It's as if my whole radar screen is lit up! I feel as if I've upgraded my brain."

Acknowledgments

WHEN I FIRST TOLD YOU, MY READERS, ABOUT THIS BOOK, I WAS MOVED by your response. Hundreds of you shared your lifelong struggles with physical and brain-related conditions, including depression, anxiety, mood disorders, learning and cognitive issues, and autoimmune conditions, as well as those of your sons, daughters, husbands, wives, sisters, brothers, parents, and friends. You also shared the disappointment, fatigue, and grief you've so often felt at not having yet found real and lasting answers to help yourself or your loved ones achieve the well-being you long for.

To all of you who wrote to me and shared your deeply honest and moving stories, a heartfelt thank-you. This book is for you, and the millions of others who suffer alongside you with intractable brain-related and mental health disorders.

To the women who allowed me to follow them for a year as they searched for answers of their own, and to tell their stories in these pages, thank you for your courage, your determination, your openness. Working with you profoundly moved me, and I learned much from you.

This book wouldn't exist without the help and vision of four extraordinary women. Several years ago, my friend the well-known neuroscientist

Peg McCarthy began telling me about a little brain cell called microglia that was changing the way she thought about the brain, and disorders of the brain. Her excitement proved contagious. As Peg—one of the most talented communicators of complex science I've ever known—answered my questions about this enigmatic little cell, over dinners and in emails, I became deeply intrigued by the possibilities that this science offered. Our exchanges over a two-year period provided the early seeds that would later grow into this book, and I thank her.

Slowly, as I began to tie the research on microglia together with our new understanding of the brain as an immune organ, and what this meant in terms of addressing patients' mental health, my lifelong agent and friend, Elizabeth Kaplan, listened (many times) as I went on about the hope this might soon offer patients. One day she picked up the phone and called my all-time favorite editor, Marnie Cochran, with whom I'd previously worked, and asked if she would simply listen as I talked through my still loosely formed concept for this book. That first conversation in the fall of 2016 lasted for two hours, and I will never forget looking out over the fields outside my attic writing studio as Marnie, who immediately understood the importance of this science, helped me determine how to tell this story: Follow the tale of this little cell, microglia, and bring the cell to life. This idea grew into a book concept that afternoon. (Soon after we hung up the phone, she also came up with the title.)

But of course this book wouldn't exist without the scientists who stand at the center of the research. I'm especially indebted to Beth Stevens for trusting me with her story, for allowing me to interview and shadow her, and for the many hours she devoted to this project. She helped me get the science right with precision, kindness, and patience.

My thanks to these four women—Peg for first getting me excited about microglia; Elizabeth for listening, believing this story mattered, and finding it a home; Marnie for being the best kind of collaborative book editor (dare I say an old-fashioned one?) who partners, coaches, and nurtures at every step of the book writing process, and by so doing, made this book so much better; and Beth, for leading the charge with this research, in ways that fundamentally promise to change our understanding of disorders of the mind and improve patients' lives.

I believe this book exists because of the power of relationships and what happens when we connect, engage in deeper conversations, and listen, with the hope of creating something new and purposeful, in the desire to help others.

My deepest gratitude too to all of the extraordinary scientists who joined in this project, especially Alan Faden, Jony Kipnis, Valter Longo, Antoine Louveau, Andrew Miller, Alvaro Pascual-Leone, Charles Raison, Dori Schafer, Li-Huei Tsai, and Susannah Tye. Thank you for sharing your powerful discoveries and ideas in these pages.

I'm grateful to practitioners Hasan Asif and Mark Trullinger for allowing me to report on their work with patients, as well as to Sebern Fisher and Jay Gunkelman for shedding light on the science of neurofeedback.

To my readers: Any scientific revelations you glean here are thanks to the work of those I've profiled in these pages; if there are any errors, they are entirely my own.

I owe an inestimable debt of gratitude to my own personal support team, my friend-family of oh so many years. Kimberly Minear, how do I thank you? Christy Bethell, Shannon Brownlee, Faith Hackett, Sarah Judd, Amy Karlen, Barbee Whitaker, Bob Whitaker, your friendship and big, kind hearts keep me sane on this planet.

To my early readers Nina Haigney, Ahmet Hoke, Sarah Judd, Amy Karlen, Peg McCarthy, Diane Petrella, thank you for giving me your important feedback and thoughts. Shannon Brownlee, your editorial suggestions significantly improved these pages; thank you for making this a better book.

The time period during which I wrote this book was a particularly challenging one in terms of my own health, and also that of several family members whom I love deeply. To the practitioners who have helped me manage my own health issues, so that I can meet deadlines, travel to deliver lectures, and caregive, thank you for reminding me to take care of my mind and body first. To Anita Bains, Anat Baniel, Marti Glenn, Jim Hill, Al Liao, Lisa Madill, Joshua Nachman, Hiroshi Nakazawa,

Georgie Oldfield, Diane Petrella, Megan Rich, Marla Sanzone, and Eric Schneider, thank you.

I also thank the Virginia Center for the Creative Arts for a glorious three-week fellowship that allowed me to work on this book in a small writing studio nestled in the Blue Ridge Mountains. That time proved invaluable.

Finally, and most important, thank you to my family. My husband, Zen, my best friend and partner through so much that cannot be written here—as you like to say, "I love how we grow together." I would never have made it through this deadline, and the challenges of our own families' health concerns, without you (not to mention your gluten-free, no-added-sugar apple pies). To my son, Christian, and daughter, Claire, who've grown up seeing me conduct endless phone interviews, jump on Amtrak hundreds of times to do face-to-face reporting, and edit thousands of pages at the kitchen table over the past twenty years, I loved you the day you were born, and I love and admire the young man and the young woman you have become.

This book is dedicated to my daughter, Claire, her fortitude, her courage, her big heart.

Notes

Prologue: When the Body Attacks the Brain

7 **In 2008, research revealed that patients with MS:** N. D. Chiaravalloti and J. DeLuca, "Cognitive Impairment in Multiple Sclerosis," *Lancet Neurology* 7 (December 2008), 1139–51. Based on later research, the American Academy of Neurology would issue a comprehensive report, in 2013, that "one-third to one-half of people with MS will have a major depressive episode" in their lifetime, as compared to less than one-fifth of the general population; that anxiety disorders "affect more than one-third of people with MS"; and that bipolar disorder occurs in 13 percent of people with MS and in less than 5 percent of people without MS. "Summary of Evidence-Based Guideline for Patients and Their Families: Emotional Disorders in People with Multiple Sclerosis," American Academy of Neurology, www.aan.com/Guidelines/Home/GetGuidelineContent/630 (accessed July 3, 2017).

7 **Shockingly, as many as 56 percent of patients with lupus:** A. Unterman, J. E. S. Nolte, M. Boaz, et al., "Neuropsychiatric Syndromes in Systemic Lupus Erythematosus: A Meta-analysis," *Seminars in Arthritis and Rheumatism* 41, no. 1 (August 14, 2011), 1–11. This study was originally published online on October 20, 2010. Researchers performed a meta-analysis of 17 separate studies looking at a total of 5,057 patients with systemic lupus erythematosus, in order to comprehensively examine the link between lupus and neuropsychiatric disorders. In a much earlier study, in 2001, a high correlation between lupus and neuropsychiatric disorders was noted in 46 patients: H. Ainiala, J. Loukkola, J. Peltola, et al., "The Prevalence of Neuropsychiatric Syndromes in Systemic Lupus Erythematosus," *Neurology* 57, no. 3 (August 2001), 496–500. In 2015, researchers presented an updated overview of studies showing a link

between autoimmune conditions and neuropsychiatric symptoms: R. Sankowski, S. Mader, and S. I. Valdés-Ferrer, "Systemic Inflammation and the Brain: Novel Roles of Genetic, Molecular, and Environmental Cues as Drivers of Neurodegeneration," *Frontiers in Cellular Neuroscience* 9 (February 2015), 1–20.

7 **Having lupus is also associated with early dementia:** Y. Shoenfeld, O. Gendelman, S. Tiosano, et al., "High Proportions of Dementia Among SLE Patients: A Big Data Analysis," *International Journal of Geriatric Psychiatry* 33, no. 3 (March 2018), 531–36.

7 **And, that same year, researchers found:** M. E. Benros, B. L. Waltoft, M. Nordentoft, et al., "Autoimmune Diseases and Severe Infections as Risk Factors for Mood Disorders: A Nationwide Study," *JAMA Psychiatry* 70, no. 8 (August 2013), 812–20. This research was conducted between 1977 and 2010 and looked at a total of 3.56 million people.

7 **In one case study, a patient who received a bone marrow transplant:** I. E. Sommer, D. W. van Bekkum, H. Klein, et al., "Severe Chronic Psychosis after Allogeneic SCT from a Schizophrenic Sibling," *Bone Marrow Transplant* 50, no. 1 (January 2015), 153–54.

7 **In another case study, a young man with schizophrenia:** T. Miyaoka, R. Wake, S. Hashioka, et al., "Remission of Psychosis in Treatment-Resistant Schizophrenia following Bone Marrow Transplantation: A Case Report," *Frontiers in Psychiatry* 8, no. 174 (September 2017), doi:10.3389/fpsyt.2017.00174.

9 **So early anatomists had very good reason:** Infections that directly attacked the brain, such as meningitis, were another exception to this "brain as immune-privileged" rule.

One: The Accidental Neurobiologist

16 **(A fellow neuroscientist once described Beth Stevens):** Emily Underwood reported this in her article, "This Woman May Know a Secret to Saving the Brain's Synapses," *Science* (August 18, 2016), www.sciencemag.org/news/2016/08/woman-may-know-secret-saving-brain-s-synapses (accessed October 29, 2017).

20 **Fields was interested in Schwann cells:** R. Douglas Fields, *The Other Brain* (New York: Simon & Schuster, 2011).

23 **Axel Nimmerjahn, Ph.D., who presented this work:** A. Nimmerjahn, F. Kirchhoff, and F. Helmchen, "Resting Microglial Cells Are Highly Dynamic Surveillants of Brain Parenchyma in Vivo," *Science* 308, no. 5726 (May 27, 2005), 1314–18.

24 **But instead of staying in the body like white blood cells:** F. Ginhoux, M. Greter, M. Leboeuf, et al., "Fate Mapping Analysis Reveals That Adult Microglia Derive from Primitive Macrophages," *Science* 330, no. 6005 (November 2010), 841–45. In their recent book, Vladimir Maletic and Charles Raison unveil how microglia are present during the earliest moment of neural

development, before other glial cells: "As a matter of fact, microglial ancestors have migrated from the yolk sack and joined the development of the neural tube, predating the arrival of the astrocyte/oligodendrocyte predecessors. . . . The early partnership between neurons and microglia bespeaks of its cardinal role in brain development." Vladimir Maletic and Charles Raison, *The New Mind-Body Science of Depression* (New York: W. W. Norton, 2017), 263.

26 **We see this in diseases like rheumatoid arthritis:** A. Laria, A. M. Lurati, M. Marrazza, et al., "The Macrophages in Rheumatic Diseases," *Journal of Inflammation Research* 9 (February 2016), 1–11.

27 **When synapses were tagged by complement:** B. Stevens, N. J. Allen, L. E. Vasquez, et al., "The Classical Complement Cascade Mediates CNS Synapse Elimination," *Cell* 131, no. 6 (December 14, 2007), 1164–78.

Two: "Ten Feet Out of a Forty-Foot Well"

35 **I am reminded of words Vincent van Gogh wrote:** On her blog, *Brain Pickings*, Maria Popova shares a quote from Vincent van Gogh's writings about depression in an essay called "What Depression Is Really Like," www.brain pickings.org/2016/02/09/depression-william-styron-darkness-visible/ (accessed November 15, 2017).

35 **The word "melancholia" first appeared in English:** William Styron, *Darkness Visible* (New York: Random House, 1990), 36–37.

36 **The term "hysteria" derived from the Greek word for the uterus:** Matt Simon, "Fantastically Wrong: The Theory of the Wandering Wombs That Drove Women to Madness," *Wired*, May 7, 2014, www.wired.com/2014/05/ fantastically-wrong-wandering-womb/ (accessed November 29, 2017). Simon draws on Helen King's essay "Once Upon a Text: Hysteria from Hippocrates," which appears in the book *Hysteria Beyond Freud* by Sander L. Gilman, Helen King, Roy Porter, G. S. Rousseau, and Elaine Showalter (Berkeley: University of California Press, 1993).

36 **Then, about a hundred years ago, a Swiss-born psychiatrist:** William Styron, *Darkness Visible* (New York: Random House, 1990), 37.

36 **The word stuck—though, as William Styron writes:** Ibid.

37 **Given advances in neuroscience, he posited:** Thomas Insel, "Transforming Diagnosis," *National Institute of Mental Health* (April 29, 2013), www.nimh .nih.gov/about/directors/thomas-insel/blog/2013/transforming-diagnosis.shtml (accessed October 9, 2017).

Three: Friendly Fire in the Brain

44 **Such synapse loss in the retina was known to lead to diseases:** B. Stevens, B. A. Barres, N. J. Allen, et al., "The Classical Complement Cascade Medi-

ates CNS Synapse Elimination," *Cell* 131, no. 6 (December 2007), 1164–78; G. R. Howell, D. G. Macalinao, G. L. Sousa, et al., "Molecular Clustering Identifies Complement and Endothelin Induction as Early Events in a Mouse Model of Glaucoma," *Journal of Clinical Investigation* 121, no. 4 (April 2011), 1429–44.

46 **It was the first scientific study offering evidence:** D. P. Schafer, E. K. Lehrman, A. G. Kautzman, et al., "Microglia Sculpt Postnatal Neural Circuits in an Activity and Complement-Dependent Manner," *Neuron* 74, no. 4 (May 2012), 691–705.

46 **Meanwhile, researchers at the European Molecular Biology Laboratory:** R. C. Paolicelli, G. Bolasco, F. Pagani, et al., "Synaptic Pruning by Microglia Is Necessary for Normal Brain Development," *Science* 333, no. 6048 (September 9, 2011), 1456–58.

48 **They even release neuroprotective factors:** Lisa Bain, Noam I. Keren, and Sheena M. Posey Norris, *Biomarkers of Neuroinflammation: Proceedings of a Workshop* (Washington, DC: National Academies Press, 2018), 18.

48 **In fact, microglia can directly help neurons:** A. Miyamoto, H. Wake, A. W. Ishikawa, et al., "Microglia Contact Induces Synapse Formation in Developing Somatosensory Cortex," *Nature Communications* (August 25, 2016), 12540.

49 **Scientists now hypothesize that once microglia are triggered:** S. J. Yu, J. W. VanRyzin, M. Perez-Pouchoulen, et al., "Temporary Depletion of Microglia during the Early Postnatal Period Induces Lasting Sex-Dependent and Sex-Independent Effects on Behavior in Rats," *eNeuro* 3, no. 6 (November–December 2016), dx.doi.org/10.1523/ENEURO.0297-16.2016.

50 **In 2016—with generous foundation funding—Stevens and her team:** B. Stevens, S. Hong, B. A. Barres, et al., "Complement and Microglia Mediate Early Synapse Loss in Alzheimer Mouse Models," *Science* 352, no. 6286 (May 2016), 712–16.

50 **In Alzheimer's, abnormally high levels:** Emily Underwood, "This Woman May Know a Secret to Saving the Brain's Synapses," *Science* (August 18, 2016), www.sciencemag.org/news/2016/08/woman-may-know-secret-saving-brain-s -synapses (accessed October 29, 2017).

51 **This aligned with the work of other researchers:** M. L. MacDonald, J. Alhassan, J. T. Newman, et al., "Selective Loss of Smaller Spines in Schizophrenia," *American Journal of Psychiatry* 174, no. 6 (June 1, 2017), 586–94; D. A. Lewis, S. J. Dienel, and H. H. Bazmi, "Development of Transcripts Regulating Dendritic Spines in Layer 3 Pyramidal Cells of the Monkey Prefrontal Cortex: Implications for the Pathogenesis of Schizophrenia," *Neurobiology of Disease* 105 (September 2017), 132–41.

51 **In 2016, a colleague of Beth's, the geneticist Steven McCarroll:** A. Sekar, A. R. Bialas, H. de Rivera, et al. "Schizophrenia Risk from Complex Variation of Complement Component 4," *Nature* 530, no. 7589 (2016), 177–83; Lisa

Bain, Noam I. Keren, and Sheena M. Posey Norris, *Biomarkers of Neuro-inflammation: Proceedings of a Workshop* (Washington, DC: National Academies Press, 2018), 19–34.

Four: Microglia Everywhere

56 **When children face chronic unpredictable stress:** J. E. Lin, T. C. Neylan, E. Epel, et al., "Association of Childhood Adversity and Adulthood Trauma with C-Reactive Protein: A Cross-Sectional Population-Based Study," *Brain, Behavior, and Immunity* 53 (March 2016), 105–12.

56 **And this in turn can reset the stress response to high:** B. Labonté, M. Suderman, G. Maussion, et al., "Genome-Wide Epigenetic Regulation by Early Life Trauma," *Archives of General Psychiatry* 69, no. 7 (July 2012), 722–31; S. E. Romens, J. McDonald, J. Svaren, et al., "Associations Between Early Life Stress and Gene Methylation in Children," *Child Development* 86, no. 1 (January/February 2015); M. J. Meaney and M. Szyf, "Environmental Programming of Stress Responses Through DNA Methylation: Life at the Interface Between a Dynamic Environment and a Fixed Genome," *Dialogues, Clinical Neuroscience* 7, no. 2 (2005), 103–23; and M. Suderman, P. O. McGowan, A. Sasaki, et al., "Conserved Epigenetic Sensitivity to Early Life Experience in the Rat and Human Hippocampus," *Proceedings of the National Academy of Sciences* 109, suppl. 2 (October 16, 2012), 17266–72.

56 **In fact, Yale researchers recently found:** N. Weder, H. Zhang, K. Jensen, et al., "Child Abuse, Depression, and Methylation in Genes Involved with Stress, Neural Plasticity, and Brain Circuitry," *Journal of the American Academy of Child and Adolescent Psychiatry* 53, no. 4 (April 2014), 417–24.

56 **And this is the reason why children who grow up:** S. R. Dube, D. Fairweather, W. S. Pearson, et al., "Cumulative Childhood Stress and Autoimmune Diseases in Adults," *Psychosomatic Medicine* 71, no. 2 (February 2009), 243–50; M. Dong, W. H. Giles, V. J. Felitti, et al., "Insights into Causal Pathways for Ischemic Heart Disease: Adverse Childhood Experiences Study," *Circulation* 110, no. 13 (September 28, 2004), 1761–66; D. W. Brown, R. F. Anda, V. J. Felitti, et al., "Adverse Childhood Experiences Are Associated with the Risk of Lung Cancer: A Prospective Cohort Study," *BioMed Central Public Health* (January 19, 2010), 20; and R. D. Goodwin and M. B. Stein, "Association Between Childhood Trauma and Physical Disorders Among Adults in the United States," *Psychological Medicine* 34, no. 3 (April 2004), 509–20. For more on the relationship between ACE scores and disease, see www.cdc.gov/ace/outcomes.htm. B. Z. Yang, H. Zhang, G. Wenjing, et al., "Child Abuse and Epigenetic Mechanisms of Disease Risk," *American Journal of Preventive Medicine* 44, no. 2 (February 2013), 101–17.

56 **We also know that children who experience a lot of childhood stress:** D. P.

Chapman, C. L. Whitfield, V. J. Felitti, et al., "Adverse Childhood Experiences and the Risk of Depressive Disorders in Adulthood," *Journal of Affective Disorders* 82, no. 2 (October 15, 2004), 217–25.

56 **When researchers run brain scans on adults:** M. A. Sheridan, N. A. Fox, C. H. Zeanah, et al., "Variation in Neural Development as a Result of Exposure to Institutionalization Early in Childhood," *Proceedings of the National Academy of Sciences* 109, no. 32 (August 7, 2012), 12927–32.

57 **When the hippocampus's circuitry is oversculpted:** L. Schmaal, D. J. Veltman, T. G. M. van Erp, et al., "Subcortical Brain Alterations in Major Depressive Disorder: Findings from the ENIGMA Major Depressive Disorder Working Group," *Molecular Psychiatry* 21, no. 6 (June 2016), 806–12.

57 **This means that kids like Jason may have changes:** R. J. Herringa, R. M. Birn, P. L. Ruttle, et al., "Childhood Maltreatment Is Associated with Altered Fear Circuitry and Increased Internalizing Symptoms by Late Adolescence," *Proceedings of the National Academy of Sciences* 110, no. 47 (November 19, 2013), 19119–24; E. R. Edmiston, F. Wang, C. M. Mazure, et al., "Corticostriatal-Limbic Gray Matter Morphology in Adolescents with Self-Reported Exposure to Childhood Maltreatment," *Archives of Pediatrics & Adolescent Medicine* 165, no. 12 (December 2011), 1069–77; and J. Czerniawski and J. F. Guzowski, "Acute Neuroinflammation Impairs Context Discrimination Memory and Disrupts Pattern Separation Processes In Hippocampus," *Journal of Neuroscience* 34, no. 37 (September 10, 2014), 12470–80.

58 **Patients like Katie, who suffer from major depressive disorder:** R. Haapakoski, J. Mathieu, K. P. Ebmeier, et al., "Cumulative Meta-Analysis of Interleukins 6 and Iβ, Tumour Necrosis Factor α and C-Reactive Protein in Patients with Major Depressive Disorder," *Brain, Behavior, and Immunity* 49 (October 2015), 206–15; and M. S. Cepeda, P. Stang, and R. Makadia, "Depression Is Associated with High Levels of C-Reactive Protein and Low Levels of Fractional Exhaled Nitric Oxide: Results from the 2007–2012 National Health and Nutrition Examination Surveys," *Journal of Clinical Psychiatry* 77, no. 12 (December 2016), 1666–71.

58 **For instance, women with elevated levels of C-reactive protein:** Lisa Bain, Noam I. Keren, and Sheena M. Posey Norris, *Biomarkers of Neuroinflammation: Proceedings of a Workshop* (Washington, DC: National Academies Press, 2018), 34.

58 **And kids who at the age of ten:** G. M. Khandaker, R. M. Pearson, P. B. Jones, et al., "Association of Serum Interleukin 6 and C-Reactive Protein in Childhood with Depression and Psychosis in Young Adult Life: A Population-Based Longitudinal Study," *JAMA Psychiatry* 71, no. 10 (October 2014), 1121–28.

58 **In 2015, researchers found higher levels of the cytokine:** R. N. Spengler, V. Fasick, S. Samankan, et al., "The Hippocampus and TNF: Common Links Between Chronic Pain and Depression," *Neuroscience & Behavioral Review* 53 (June 2015), 139–59.

58 **In patients with bipolar disorder, inflammatory biomarkers skyrocket:** E. Brietzke, L. Sterts, B. S. Fernandes, et al., "Comparison of Cytokine Levels in Depressed, Manic and Euthymic Patients with Bipolar Disorder," *Journal of Affective Disorders* 116, no. 3 (August 2009), 214–17; H. Yamamori, T. Ishima, Y. Yasuda, et al., "Assessment of a Multi-Assay Biological Diagnostic Test for Mood Disorders in a Japanese Population," *Neuroscience Letters* 612 (January 26, 2016), 167–71; F. Dickerson, E. Katsafanas, L. A. Schweinfurth, et al., "Immune Alterations in Acute Bipolar Depression," *Acta Psychiatrica Scandinavica* 132, no. 3 (September 2015), 204–10.

58 **Neuropsychiatrists have also found relationships:** R. Hou, M. Garner, C. Holmes, et al., "Peripheral Inflammatory Cytokines and Immune Balance in Generalised Anxiety Disorder: Case-Controlled Study," *Brain, Behavior, and Immunity* 62 (May 2017), 212–18; H. Engler, P. Brendt, J. Wischermann, et al., "Selective Increase of Cerebrospinal Fluid IL-6 During Experimental Systemic Inflammation in Humans: Association with Depressive Symptoms," *Molecular Psychiatry* 22 (October 2017), 1448–54; A. H. Miller and C. L. Raison, "The Role of Inflammation in Depression: From Evolutionary Imperative to Modern Treatment Target," *Nature Reviews: Immunology* 16 (December 2015), 22–34; M. B. Howren, D. M. Lamkin, and J. Suls, "Associations of Depression with C-Reactive Protein, IL-1, and IL-6: A Meta-Analysis," *Psychosomatic Medicine* 71, no. 2 (February 2009), 171–86; Y. Dowlati, N. Herrmann, W. Swardfager, et al., "A Meta-Analysis of Cytokines in Major Depression," *Biological Psychiatry* 67 no. 5 (March 1, 2010), 446–57; Lisa Bain, Noam I. Keren, and Sheena M. Posey Norris, *Biomarkers of Neuroinflammation: Proceedings of a Workshop* (Washington, DC: National Academies Press, 2018), 25.

58 **This is also the case in schizophrenia:** F. P. Hartwig, M. C. Borges, B. L. Horta, et al., "Inflammatory Biomarkers and Risk of Schizophrenia: A 2-Sample Mendelian Randomization Study," *JAMA Psychiatry* 74, no. 12 (December 2017), 1226–33.

58 **Moreover, this turns out to be true even when no signs:** C. L. Raison, C. A. Lowry, G.A.W. Rook, "Inflammation, Sanitation and Consternation: Loss of Contact with Co-Evolved, Tolerogenic Micro-Organisms and the Pathophysiology and Treatment of Major Depression," *Archives of General Psychiatry* 67, no. 12 (December 2010), 1211–24.

58 **Perhaps most astonishing, in 2017, researchers at Johns Hopkins:** F. Dickerson, H. C. Wilcox, M. Adamos, et al., "Suicide Attempts and Markers of Immune Response in Individuals with Serious Mental Illness," *Journal of Psychiatric Research* 87 (April 2017), 37–43; F. Dickerson, M. Adamos, E. Katsafanas, et al., "The Association Between Immune Markers and Recent Suicide Attempts in Patients with Serious Mental Illness: A Pilot Study," *Psychiatry Research* 255 (September 2017), 8–12.

58 **Simply activating an individual's inflammatory immune response:** Moises

Velasquez-Manoff, "When the Body Attacks the Mind," *The Atlantic* (July/August 2016), www.theatlantic.com/magazine/archive/2016/07/when-the-body-attacks-the-mind/485564/ (accessed October 12, 2017).

58 **Individuals with autism have elevated inflammatory immune markers:** G. B. Choi, Y. S. Yim, H. Wong, et al., "The Maternal Interleukin-17a Pathway in Mice Promotes Autism-Like Phenotypes in Offspring," *Science* 351, no. 6276 (February 26, 2016), 933–39.

59 **This process would take a lot longer in humans:** T. Kreisel, M. G. Frank, T. Licht, et al., "Dynamic Microglial Alterations Underlie Stress-Induced Depressive-Like Behavior and Suppressed Neurogenesis," *Molecular Psychiatry* 19, no. 6 (June 2014), 699–709.

59 **Indeed, scientists now believe that most molecular pathways to depression:** Ibid.

59 **And periods of worsened anxiety:** G. Singhal and B. T. Baune, "Microglia: An Interface Between the Loss of Neuroplasticity and Depression," *Frontiers in Cellular Neuroscience* 11 (September 8, 2017), 270.

59 **In a recent study in *JAMA Psychiatry*:** E. Setiawan, A. A. Wilson, R. Mizrahi, et al., "Role of Translocator Protein Density, a Marker of Neuroinflammation, in the Brain During Major Depressive Episodes," *JAMA Psychiatry* 72, no. 3 (March 2015), 268–75.

59 **And patients with obsessive-compulsive disorder:** S. Attwells, E. Setiawan, A. A. Wilson, et al., "Inflammation in the Neurocircuitry of Obsessive-Compulsive Disorder," *JAMA Psychiatry* 74, no. 8 (August 2017), 833–40.

59 **A similar finding was announced in 2017:** A. R. Bialas, J. Presumey, A. Das, et al., "Microglia-Dependent Synapse Loss in Type 1 Interferon-Mediated Lupus," *Nature* 546, no. 7659 (June 22, 2017), 539–43.

60 **Likewise, patients with MS who report cognitive and memory issues:** C. F. Lucchinetti, F. G. Bogdan, B. F. Popescu, et al., "Inflammatory Cortical Demyelination in Early Multiple Sclerosis," *New England Journal of Medicine* 365, no. 23 (December 8, 2011), 2188–97; R. Sankowski, S. Mader, and S. I. Valdes-Ferrer, "Systemic Inflammation and the Brain: Novel Roles of Genetic, Molecular, and Environmental Cues as Drivers of Neurodegeneration," *Frontiers in Cellular Neuroscience* 9 (February 2, 2015), 28.

60 **Meanwhile, the reason why patients with Crohn's disease:** D. R. van Langenberg, G. W. Yelland, S. R. Robinson, et al., "Cognitive Impairment in Crohn's Disease Is Associated with Systemic Inflammation, Symptom Burden and Sleep Disturbance," *United European Gastroenterology Journal* 5, no. 4 (June 2017), 579–87.

60 **Somehow, the body's hyperactive immune system:** Ibid. The study's authors note, "While the basis of this effect is not fully understood, experiments on rodents have shown that inflammation of the colon results in an upregulation of inflammatory activity in microglia of the hippocampus that in turn causes profound impairments in post-synaptic responses. If such decrements in post-synaptic responsiveness occur in humans with CD, they could readily account

for the slower response times observed in the present study." Also see K. Riazi, M. A. Galic, A. C. Kentner, et al., "Microglia-Dependent Alteration of Gluamatergic Synaptic Transmission and Plasticity in the Hippocampus During Peripheral Inflammation," *Journal of Neuroscience* 35, no. 12 (March 25, 2015), 4942–52.

60 **PET scans of the brains of men with autism:** Virginia Hughes, "Brain Imaging Study Points to Microglia as Autism Biomarker," *Spectrum*, January 10, 2013, spectrumnews.org/news/brain-imaging-study-points-to-microglia-as-autism -biomarker/ (accessed October 12, 2017); and J. L. Frost and D. P. Schafer, "Microglia: Architects of the Developing Nervous System," *Trends in Cell Biology* 26, no. 8 (August 1, 2016), 587–96.

60 **In individuals with autism, microglia:** S. Katsuaki, G. Sugihara, Y. Ouchi, et al., "Microglial Activation in Young Adults with Autism Spectrum Disorder," *JAMA Psychiatry* 70, no. 1 (January 2013); Y. Mizoguchi and A. Monji, "Microglial Intracellular Ca^{2+} Signaling in Synaptic Development and Its Alterations in Neurodevelopmental Disorders," *Frontiers in Cellular Neuroscience* 11 (March 17, 2017), 69; and S. Gupta, S. E. Ellis, F. N. Ashar, et al., "Transcriptome Analysis Reveals Dysregulation of Innate Immune Response Genes and Neuronal Activity-Dependent Genes in Autism," *Nature Communications* 5 (December 2014), 5748.

60 **Microglia also promote disease progression:** S. R. Subramaniam and H. J. Federoff, "Targeting Microglial Activation States as a Therapeutic Avenue in Parkinson's Disease," *Frontiers in Aging Neuroscience* 9 (June 2017), 176.

60 **even in West Nile virus:** M. J. Vasek, C. Garber, D. Dorsey, et al., "A Complement-Microglial Axis Drives Synapse Loss during Virus-Induced Memory Impairment," *Nature* 534, no. 7608 (June 2016), 538–43.

63 **The field of neuroimmunology helps us to better grasp:** W. W. Eaton, M. G. Pedersen, P. R. Nielsen, et al., "Autoimmune Diseases, Bipolar Disorder, and Non-Affective Psychosis," *Bipolar Disorders* 12, no. 6 (September 2010), 638–46.

64 **And those with depression were more likely:** J. Euesden, A. Danese, C. M. Lewis, et al., "A Bidirectional Relationship Between Depression and the Autoimmune Disorders: New Perspectives from the National Child Development Study," *PLOS One* 12, no. 3 (March 6, 2017), e0173015.

65 **As the scientific philosopher Thomas Kuhn:** The twentieth-century scientific philosopher Thomas Kuhn referred to this lag time between scientific discovery and a paradigm shift occurring in scientific understanding "normal science." Science, Kuhn argued, is conservative in nature, unwilling to abandon ideas without persuasive evidence. The overwhelming majority of scientists accept a single scientific ideology as the starting point from which they form their own viewpoint—the pathway from which they view the entire scientific landscape—to the degree that they cannot overturn that ideology, even if research begins to show that it is blatantly leading them astray.

Five: A Bridge to the Brain

66 **It has sometimes been said that we know:** Matt Haig, "Kurt Cobain Was Not a 'Tortured Genius,' He Had an Illness," *The Telegraph*, April 5, 2015. In this article, Haig writes, "Neuroscience is a baby science, a mere century old, and our scientific understanding of the brain is nowhere near where we'd like it to be. We know more about the moons of Jupiter than what is inside of our skulls." You can read his excellent article here: www.telegraph.co.uk/men/ thinking-man/11515605/Kurt-Cobain-was-not-a-tortured-genius-he-had-an -illness.html (accessed November 5, 2017).

67 **Conversely, when he reintroduced:** J. Kipnis, H. Cohen, M. Cardon, et al., "T-Cell Deficiency Leads to Cognitive Dysfunction: Implications for Therapeutic Vaccination of Schizophrenia and Psychiatric Conditions," *Proceedings of the National Academy of Sciences* 101, no. 21 (May 2004), 8180–85.

67 **He wondered whether the body's T-cells:** I. Shaked, Z. Porat, R. Gersner, et al., "Early Activation of Microglia as Antigen-Presenting Cells Correlates with T Cell–Mediated Protection and Repair of the Injured Central Nervous System," *Journal of Neuroimmunology* 146, no. 1–2 (January 2004), 84–93.

67 **In 2010, he published work showing:** N. C. Derecki, A. N. Cardani, C. H. Yang, et al., "Regulation of Learning and Memory by Meningeal Immunity: A Key Role for IL-4," *Journal of Experimental Medicine* 207, no. 5 (May 2010), 1067–80. Earlier, in 2006, Kipnis had also published a paper investigating whether it was possible, given accumulating evidence suggesting that neurodegenerative diseases of the central nervous system were also associated with inflammation, that all neurodegenerative diseases might even be associated with abnormal immunity. M. Schwartz, O. Butovsky, and J. Kipnis, "Does Inflammation in an Autoimmune Disease Differ from Inflammation in Neurodegenerative Diseases? Possible Implications for Therapy," *Journal of Neuroimmune Pharmacology* 1, no. 1 (March 2006), 4–10.

70 **But Louveau was staring down at an array of lymphatic vessels:** These experiments happened over the course of several months and it was only after multiple attempts, and staining for lymphatic vessels repeatedly, that the team realized that these were indeed lymphatic vessels.

70 **Scientists use fluorescent markers to make protein molecules:** In this section, I use the term "markers"; however, Kipnis and his colleagues use the term "reagents" to refer to these markers.

70 **They were irrefutably there:** Jonathan Kipnis, "The Seventh Sense," *Scientific American*, August 2018, 29–35. A. Louveau, I. Smirnov, J. Keyes, et al., "Structural and Functional Features of Central Nervous System Lymphatics," *Nature* 523, no. 7560 (July 16, 2015), 337–41.

71 **The next year, Kipnis's lab, in collaboration with a group:** M. Absinta, S. K. Ha, G. Nair, et al., "Human and Nonhuman Primate Meninges Harbor Lymphatic Vessels That Can Be Visualized Noninvasively by MRI," *eLife* 6 (Octo-

ber 3, 2017), e29738. Investigators were able to show that these lymphatic vessels were present in both monkey and human brains.

74 **back-and-forth messaging through other portals:** N. Lou, T. Takano, Y. Pei, et al., "Purinergic Receptor P2RY12-Dependent Microglial Closure of the Injured Blood-Brain Barrier," *Proceedings of the National Academy of Sciences* 113, no. 4 (January 26, 2016), 1074–79.

74 **There are still many questions to answer:** A. Louveau, B. A. Plog, S. Antila, et al., "Understanding the Functions and Relationships of the Glymphatic System and Meningeal Lymphatics," *Journal of Clinical Investigation* 127, no. 9 (September 2017), 3210–19.

Six: "It Seems There Are No New Solutions"

83 **In one recent year, one in six adolescent girls:** R. Mojtabai, M. Olfson, and B. Han, "National Trends in the Prevalence and Treatment of Depression in Adolescents and Young Adults," *Pediatrics* 138, no. 6 (December 2016), e20161878.

83 **In a study of one hundred thousand children:** J. Breslau, S. E. Gilman, B. D. Stein, et al., "Sex Differences in Recent First-Onset Depression in an Epidemiological Sample of Adolescents," *Translational Psychiatry* 7, no. 5 (May 2017), e1139.

83 **By the time teens hit the age of seventeen:** Ibid.

83 **According to the National Institutes of Health:** "Major Depression: Prevalence of Major Depressive Episode Among Adolescents," National Institute of Mental Health, www.nimh.nih.gov/health/statistics/prevalence/major-depression -among-adolescents.shtml (accessed November 6, 2017).

84 **In fact, 535 pediatricians from small-town:** "Rising Mental Health Issues Facing Our Children, in Five Charts," *The Atlantic,* 2014, www.theatlantic .com/sponsored/athena-where-does-it-hurt/ (accessed November 6, 2017).

85 **Between 2010 and 2013 the rate of children:** Ibid.

85 **The study's authors write, "We'd noticed":** G. Plemmons, "Trends in Suicidality and Serious Self-Harm for Children 5–17 Years at 32 U.S. Children's Hospitals, 2008–2015," presented on May 7, 2017, at the 2017 Pediatric Academic Societies Meeting in San Francisco. The United States is hardly alone in this trend; in the past six years, in Canada, there has been a one-third rise in emergency room visits for children and youth presenting with mental health and addiction problems. S. Gandhi, M. Chiu, K. Lam, et al., "Mental Health Service Use Among Children and Youth in Ontario: Population-Based Trends over Time," *Canadian Journal of Psychiatry* 61, no. 2 (February 2016), 119–24. In just the past five years, teenage suicide attempts increased 23 percent, and the number of teens aged 13 to 18 years who committed suicide jumped 31 percent. J. M. Twenge, T. E. Joiner, M. L. Rogers, et al., "Increases in De-

pressive Symptoms, Suicide-Related Outcomes, Suicide Rates Among U.S. Adolescents After 2010 and Links to Increased New Media Screen Time," *Clinical Psychological Science* (published online November 14, 2017), 1–15. Suicide has also become the second leading cause of death among both iGen and millennials between the ages of 15 and 34: "Suicide Is a Leading Cause of Death in the United States," National Institute of Mental Health, www.nimh .nih.gov/health/statistics/suicide/index.shtml#part_154968 (accessed November 6, 2017).

85 **It's not just teens and children who are suffering:** A. Case and A. Deaton, "Rising Morbidity and Mortality in Midlife Among White Non-Hispanic Americans in the 21st Century," *Proceedings of the National Academy of Sciences* 112, no. 49 (December 2015), 15078–83. Midlife increases in depression, suicide, and drug overdose have been driving a rise in midlife mortality rates, write these two Princeton researchers: "This paper documents a marked increase in the all-cause mortality of middle-aged white non-Hispanic men and women in the United States between 1999 and 2013. This change reversed decades of progress in mortality and was unique to the United States; no other rich country saw a similar turnaround. . . . Self-reported declines in health, mental health, and ability to conduct activities of daily living, and increases in chronic pain and inability to work, as well as clinically measured deteriorations in liver function, all point to growing distress in this population. . . . If the white mortality rate for ages 45–54 had held at their 1998 value, 96,000 deaths would have been avoided from 1999–2013, 7,000 in 2013 alone. If it had continued to decline at its previous (1979–1998) rate, half a million deaths would have been avoided in the period 1999–2013. . . . Concurrent declines in self-reported health, mental health, and ability to work, increased reports of pain, and deteriorating measures of liver function all point to increasing midlife distress." Today, major depressive disorder is the most common form of disability worldwide. "Data on Behavioral Health in the United States," American Psychological Association, www.apa.org/helpcenter/ data-behavioral-health.aspx (accessed November 6, 2017). Of course, rising rates of mental health disorders aren't just an American problem: major depressive disorder ranks in the top ten causes of ill health in all but four countries worldwide. C. J. L. Murray and A. D. Lopez, "Measuring Global Health: Motivation and Evolution of the Global Burden of Disease Study," *Lancet* 390, no. 10100 (September 16, 2017), 1460–64. In England, at any given time, a sixth of the population between the ages of 16 and 64 now has a diagnosed mental health problem. David Brown and Nick Triggle, "Mental Health: 10 Charts on the Scale of the Problem," BBC News, September 30, 2017, www.bbc.com/news/health-41125009 (accessed November 6, 2017). And in 2016, 1.1 billion people around the world were living with psychological or psychiatric disorders and substance abuse problems.

85 **Rates of adult suicide have surged:** J. A. Phillips, A. V. Robin, C. N. Nugent, et al., "Understanding Recent Changes in Suicide Rates Among the Middle-

aged: Period or Cohort Effects?" *Public Health Report* 125, no. 5 (September–October 2010), 680–88. According to the Centers for Disease Control and Prevention (CDC), in 2015, suicide was the tenth leading cause of death in the United States, claiming 44,000 people each year. Or, put another way, more than 100 Americans commit suicide every day. See reporter Jack Rodolico's radio piece "Heroin Diaries," which aired on the radio show *Reveal* on October 26, 2017, and shared recent statistics on heroin and opioid deaths in the United States, www.revealnews.org/episodes/heroin-diaries/ (accessed December 6, 2017).

85 **In 2017 alone, fifty-two thousand Americans died of heroin overdoses:** Ibid.

Seven: A Modern Braindemic

89 **Let's pretend that it's five hundred years ago:** S. Gagneux, "Host-Pathogen Coevolution in Human Tuberculosis," *Philosophical Translations of the Royal Society of London, Biological Sciences* 364, no. 159 (March 19, 2012), 850–59.

89 **You want to curl up into a little ball:** G. M. Slavich and M. R. Irwin, "From Stress to Inflammation and Major Depressive Disorder: A Social Signal Transduction Theory of Depression," *Psychological Bulletin* 140, no. 3 (May 2014), 774–815.

91 **Our ancestors' immune systems became "highly educated":** Jonathan Kipnis, Ph.D., has been looking at this link between the immune system and social behavior. He hypothesizes that, in evolutionary terms, as human beings became more social, we began to mount a stronger, more robust immune response to keep us safe from all the contagious pathogens we pass along to each other in close proximity. This served a biological imperative; men and women needed to socialize together and form bonds in order to have children and allow the human race to march onward. Ingeniously, our brains and microbes evolved in a way to allow us to be both more social and more protected from pathogens—by upping our general everyday cytokine response to microbes and pathogens so we could survive being in more social situations more of the time, fall in lust and/or love, and have more babies. (This higher base level of an inflammatory immune response is very different from the much more hyped-up cytokine spike we see during times of illness, or in response to threats, and in depression.) This means that too little immune response can also be problematic. For instance, Kipnis recently found that when you remove particular cytokine messages from the meningeal spaces, mice lose all sociability. Patterns of neural circuitry in the frontal lobe area of the brain, an area that's responsible for everyday behaviors and personality, change. These turn out to be similar to circuitry changes we see in individuals with autism. The messages that the body's immune system sends to the brain must be in a pitch-perfect normal range: there can't be too many, and there also can't be too few. The brain has to get it just right—and pathogens have helped us fig-

ure out, over many centuries, how to do that. As Kipnis opines, "Part of our personality may actually be dictated by our immune system." A. J. Filiano, Y. Xu, N. J. Tustison, et al., "Unexpected Role of Interferon-γ in Regulating Neuronal Connectivity and Social Behaviour," *Nature* 535, no. 7612 (July 12, 2016), 425–29.

91 **In animal studies, when newborn rats are exposed:** L. L. Williamson, E. A. McKenney, Z. E. Holzknecht, et al., "Got Worms? Perinatal Exposure to Helminths Prevents Persistent Immune Sensitization and Cognitive Dysfunction Induced by Early-Life Infection," *Brain, Behavior, and Immunity* 51 (January 2016), 14–28.

92 **Nevertheless, all these chemicals are EPA-approved:** For more on this, see my book *The Autoimmune Epidemic: Bodies Gone Haywire in a World out of Balance* (New York: Touchstone, 2008).

95 **You'd be exposed to the elements, predators:** C. L. Raison and A. H. Miller, "Pathogen-Host Defense in the Evolution of Depression: Insights into Epidemiology, Genetics, Bioregional Differences and Female Preponderance," *Neuropsychopharmacology* 42, no. 1 (January 2017), 5–27.

96 **Raison points out that many of these microorganisms:** M. G. Frank, L. K. Fonken, S. D. Dolzani, et al., "Immunization With *Mycobacterium Vaccae* Induces an Anti-Inflammatory Milieu in the CNS: Attenuation of Stress-Induced Microglial Priming, Alarmins and Anxiety-Like Behavior," *Brain, Behavior, and Immunity* 73 (October 2018), 352–63.

96 **Our confused immune system sees these constant twenty-first-century:** A. H. Miller and C. L. Raison, "The Role of Inflammation in Depression: From Evolutionary Imperative to Modern Treatment Target," *Nature Reviews: Immunology* 16 (January 2016), 22–34.

97 **But in today's world, social stressors:** You can read more about the way in which genes influencing depression also bolstered our ancestors' immune systems in Brian Gabriel's piece "The Evolutionary Advantage of Depression," *The Atlantic*, October 2, 2012.

98 **All of these social stressors, experts universally argue:** Michael Ian Black's recent op-ed, "The Boys Are Not All Right," *New York Times*, February 21, 2018, does not directly address the effect of social stressors on the developing male brain; however, Black powerfully argues that today's mass school shootings are a sign that young men today are not okay, that they are trapped in "the same suffocating, outdated model of masculinity, where manhood is measured in strength, where there is no way to be vulnerable without being emasculated, where manliness is about having power over others." This, coupled with a lack of understanding of how to express grief, fear, vulnerability, or tenderness, leaves men with a choice between withdrawal and rage. This is certainly, I would argue, a social stressor on boys' developing brains.

99 **Trauma researchers who study the effects of adverse:** This is based on my conversation with Jane Stevens, founder and editor of ACEsTooHigh and ACEsConnection, news sites funded by the Robert Wood Johnson Founda-

tion. On November 11, 2017, Stevens, a world-renowned expert on adverse childhood experiences, confirmed for me that the use of social media is now considered an adverse childhood experience.

101 **Not surprisingly, Johns Hopkins researchers:** R. Mojtabai, M. Olfson, and B. Han, "National Trends in the Prevalence and Treatment of Depression in Adolescents and Young Adults," *Pediatrics* 138, no. 6 (December 2016), e20161878.

101 **In fact, the use of Facebook and other social media sites:** E. Kross, P. Verduyn, E. Demiralp, et al., "Facebook Use Predicts Declines in Subjective Well-Being in Young Adults," *PLOS One* 8, no. 8 (August 2013), e69841; Holly B. Shakya and Nicholas A. Christakis, "A New, More Rigorous Study Confirms: The More You Use Facebook, the Worse You Feel," *Harvard Business Review*, April 10, 2017.

101 **Teens who spend more than five hours a day online:** J. M. Twenge, T. E. Joiner, M. L. Rogers, et al., "Increases in Depressive Symptoms, Suicide Rates Among U.S. Adolescents After 2010 and Links to Increased New Media Screen Time," *Clinical Psychological Science* (published online November 14, 2017), 1–15.

101 **There are stronger associations between social media use:** Y. Kelly, A. Zilanawala, C. Booker, et. al., "Social Media Use and Adolescent Mental Health: Findings from the UK Millennium Cohort Study," *EClinicalMedicine* (December 2018), 59–68.

101 **In 2012, 50 percent of Americans had a smartphone:** Aaron Smith, "Record Shares of Americans Now Own Smartphones, Have Home Broadband," *Fact Tank,* Pew Research Center, January 12, 2017, www.pewresearch.org/fact-tank/2017/01/12/evolution-of-technology/ (accessed December 7, 2017).

101 **By 2015, 73 percent of teenagers had smartphones:** Pew Research Center, "73% of Teens Have Access to a Smartphone; 15% Have Only a Basic Phone," Amanda Lenhart, *Teens, Social Media & Technology Overview 2015*, April 8, 2015, www.pewinternet.org/2015/04/09/teens-social-media-technology-2015/pi_2015-04-09_teensandtech_06/ (accessed December 7, 2017).

101 **Researchers found that depression:** J. M. Twenge, A. B. Cooper, T. E. Joiner, et al., "Age, Period, and Cohort Trends in Mood Disorder Indicators and Suicide-Related Outcomes in a Nationally Representative Dataset, 2005–2017," *Journal of Abnormal Psychology* 128, no. 3 (April 2019), 185–99.

Eight: Brain Hacking

103 **It is as if the artist has captured:** These paintings are by Greg Dunn, Ph.D., a neuroscientist-artist who found that the patterns of branching neurons he saw through the microscope reminded him of the aesthetic principles of Asian art. Dunn paints in the sumi-e (ink wash painting) style.

103 **And indeed, once we start to wrap our minds:** R. Masgrau, C. Guaza, R. M.

Ransohoff, et al., "Should We Stop Saying 'Glia' and 'Neuroinflammation'?" *Trends in Molecular Medicine* 23, no. 6 (June 2017), 486–500.

108 **Today, evidence increasingly indicates:** B. A. van der Kolk, H. Hodgdon, M. Gapen, et al., "A Randomized Controlled Study of Neurofeedback for Chronic PTSD," *PLOS One* 11, no. 12 (December 2016), e0166752; F. Minder, A. Zuberer, D. Brandeis, et al., "Informant-Related Effects of Neurofeedback and Cognitive Training in Children with ADHD Including a Waiting Control Phase: A Randomized-Controlled Trial," *European Child & Adolescent Psychiatry* 27, no. 8 (August 2018), 1055–66; and S. Banerjee and C. Argáez, "Neurofeedback and Biofeedback for Mood and Anxiety Disorders: A Review of Clinical Effectiveness and Guidelines," *CADTH Rapid Response Reports*, Canadian Agency for Drugs and Technologies in Health (November 2017).

109 **In one 2018 single-blind randomized controlled trial of patients with major depressive disorder:** D.M.A. Mehler, M. O. Sokunbi, I. Habes, et al., "Targeting the Affective Brain—A Randomized Controlled Trial of Real-Time fMRI Neurofeedback in Patients with Depression," *Neuropsychopharmacology* 43, no. 13 (December 2018), 2578–85.

110 **Multiple randomized clinical trials have backed up the efficacy of TMS:** Y. Levkovitz, M. Isserles, F. Padberg, et al., "Efficacy and Safety of Deep Transcranial Magnetic Stimulation for Major Depression: A Prospective Multicenter Randomized Controlled Trial," *World Psychiatry* 14, no. 1 (February 2015), 64–73; M. A. Demitrack and M. E. Thase, "Clinical Significance of Transcranial Magnetic Stimulation (TMS) in the Treatment of Pharmacoresistant Depression: Synthesis of Recent Data," *Psychopharmacology Bulletin* 42, no. 2 (2009), 5–38; and Y. Levkovitz, E. V. Harel, Y. Roth, et al., "Deep Transcranial Magnetic Stimulation Over the Prefrontal Cortex: Evaluation of Antidepressant and Cognitive Effects in Depressive Patients," *Brain Stimulation* 2, no. 4 (October 2009), 188–200.

110 **Alvaro Pascual-Leone, M.D., Ph.D., from Beth Israel Deaconess Medical Center:** T. Perera, M. S. George, G. Grammer, et al., "The Clinical TMS Society Consensus Review and Treatment Recommendations for TMS Therapy for Major Depressive Disorder," *Brain Stimulation* 9, no. 3 (May 2016), 336–46; A. Pascual-Leone, A. Valero-Cabre, J. Amengual, et al., "Transcranial Magnetic Stimulation in Basic and Clinical Neuroscience: A Comprehensive Review of Fundamental Principles and Novel Insights," *Neuroscience & Biobehavioral Reviews* 83, no. 17 (December 2017), 381-404.

110 **Some patients who experienced significant relief:** M. S. Kelly, A. J. Oliveira-Maria, M. Bernstein, et al., "Initial Response to Transcranial Magnetic Stimulation Treatment for Depression Predicts Subsequent Response," *Journal of Neuropsychiatry and Clinical Neurosciences* 29, no. 2 (Spring 2017), 179–82. Pascual-Leone also found that a patient's initial positive response to TMS could predict how greatly the patient might benefit from the typical fourteen to thirty sessions that made up a course of treatment. By 2017, more research

amassed, showing that TMS could precisely target networks relevant in depression, which in turn helped patients recover. M. J. Dubin, C. Liston, M. A. Avissar, et al., "Network-Guided Transcranial Magnetic Stimulation for Depression," *Current Behavioral Neuroscience Reports* 4, no. 1 (March 2017), 70–77.

110 **Alvaro Pascual-Leone conducted his first:** A. Pascual-Leone, B. Rubio, M. D. Catala, et al., "Rapid-Rate Transcranial Magnetic Stimulation of Left Dorsolateral Prefrontal Cortex in Drug-Resistant Depression," *Lancet* 348, no. 9022 (July 1996), 233–37.

111 **ECT, or electroconvulsive therapy:** A. Sartorius, L. Kranaster, C. Hoyer, et al., "Antidepressant Efficacy of Electroconvulsive Therapy Is Associated with a Reduction of the Innate Cellular Immune Activity in the Cerebrospinal Fluid in Patients with Depression," *World Journal of Biological Psychiatry* 19, no. 5 (August 2018), 379–89.

111 **In another 2018 study, researchers found that patients with treatment-resistant depression:** J. L. Kruse, E. Congdon, R. Olmstead, et al., "Inflammation and Improvement of Depression Following Electroconvulsive Therapy in Treatment-Resistant Depression," *Journal of Clinical Psychiatry* 79, no. 2 (March/April 2018).

111 **In 2016, researchers found that in animal studies:** N. Rimmerman, M. Abargil, L. Cohen, et al., "Microglia Mediate the Anti-Depressive Effects of Electroconvulsive Shock Therapy in Mice Exposed to Chronic Unpredictable Stress," *Brain, Behavior, and Immunity* 57, suppl. (October 2016), e20.

113 **In 2016, researchers at the University of California:** J. L. Gallant, A. G. Huth, W. A. de Heer, et al., "Natural Speech Reveals the Semantic Maps That Tile Human Cerebral Cortex," *Nature* 532 (April 2016), 453–58. Also in 2016, the Human Connectome Project also offered researchers the most comprehensive map of the brain's cerebral cortex ever made—delineating 97 new subareas that appeared structurally differently in varied times in development, and in different diseases. M. F. Glasser, T. S. Coalson, E. C. Robinson, et al., "A Multi-Modal Parcellation of Human Cerebral Cortex," *Nature* 536, no. 7615 (August 11, 2016), 171–78.

113 **This helpless behavior was correlated:** Y. Kim, Z. Perova, M. M. Mirrione, et al., "Whole-Brain Mapping of Neuronal Activity in the Learned Helplessness Model of Depression," *Frontiers in Neural Circuits* 10, no. 3 (February 3, 2016), eCollection 2016.

113 **Patients with anhedonia, who were unable:** C. B. Young, T. Chen, J. Keller, et al., "Anhedonia and General Distress Show Dissociable Ventromedial Prefrontal Cortex Connectivity in Major Depressive Disorder," *Translational Psychiatry* 6, no. 5 (May 2016), e810.

114 **Four years later, many of these patients' psychiatric disorders:** R. Nusslock, E. Harmon-Jones, L. B. Alloy, et al., "Elevated Left Mid-Frontal Cortical Activity Prospectively Predicts Conversion to Bipolar 1 Disorder," *Journal of Abnormal Psychology* 121, no. 3 (August 2012), 592–601, and R. Nusslock,

K. Walden, and E. Harmon-Jones, "Asymmetrical Frontal Cortical Activity Associated with Differential Risk for Mood and Anxiety Disorder Symptoms: An RDoC Perspective," *International Journal of Psychophysiology* 98, no. 2, pt. 2 (November 2015), 249–61. Pascual-Leone used functional magnetic resonance imaging (fMRI) to distinguish four specific subtypes of depression, based on where dysfunctional brain connectivity occurred. Clustering patients on this basis, he believed, not only helped with diagnosis, it also helped predict whether patients would respond to TMS. A. T. Drysdale, L. Grosenick, J. Downar, et al., "Resting-State Connectivity Biomarkers Define Neurophysiological Subtypes of Depression," *Nature Medicine* 23, no. 1 (January 2017), 28–38. Researchers found specific differences in the brain's limbic area as well as in the frontostriatal networks, an area of neural pathways that connect frontal lobe regions with the basal ganglia, which mediate motor, cognitive, and behavioral functions within the brain. Other researchers found that neurophysiologic measures of cortical excitability could be used as a kind of biomarker to predict which patients would best benefit from transcranial magnetic stimulation, and that changes in cortical excitability were related to different outcomes of TMS treatment in patients with major depressive disorder: B. Kobyashi, I. A. Cook, A. M. Hunter, et al., "Can Neurophysiologic Measures Serve as Biomarkers for the Efficacy of Repetitive Transcranial Magnetic Stimulation Treatment of Major Depressive Disorder?" *International Review of Psychiatry* 29, no. 2 (April 2017), 98–114.

114 **At the Mayo Clinic, TMS treatments:** S. H. Ameis, Z. J. Daskalakis, P. Szatmari, et al., "Repetitive Transcranial Magnetic Stimulation for the Treatment of Executive Function Deficits in Autism Spectrum Disorder: Clinical Trial Approach," *Journal of Child and Adolescent Psychopharmacology* 27, no. 5 (June 2017), 413–21. This pioneering work on TMS in patients with autism, done by Alvaro Pascual-Leone and Lindsay Oberman with the support of the National Institutes of Health, has since been popularized in John Elder Robison's book *Switched On* (New York: Spiegel & Grau, 2016).

114 **TMS has been shown to be effective in helping patients who suffer from obesity:** Livio Luzi, M.D., professor and head of endocrinology at the IRCCS Policlinico San Donato and the University of Milan in Italy, presented these findings at the Endocrine Society's 99th annual meeting in Orlando, Florida, on April 3, 2017. The presentation was titled "Deep Transcranial Magnetic Stimulation (dTMS) Exerts Anti-Obesity Effects via Microbiota Modulation."

115 **And the newest research on microglia shows:** C. L. Cunningham, V. Martinez-Cerdeno, and S. C. Noctor, "Microglia Regulate the Number of Neural Precursor Cells in the Developing Cerebral Cortex," *Journal of Neuroscience* 33, no. 10 (March 6, 2013), 4216–33; A. Sierra, S. Beccari, I. Diaz-Aparicio, et al., "Surveillance, Phagocytosis, and Inflammation: How Never-Resting Microglia Influence Adult Hippocampal Neurogenesis," *Neural Plasticity* (2014), 610343.

116 **Research shows that in addition:** H. F. Iaccarino, A. C. Singer, A. J. Martorell, et al., "Gamma Frequency Entrainment Attenuates Amyloid Load and Modifies Microglia," *Nature* 540, no. 7632 (December 7, 2016), 230–35.

116 **Any deviation from the norm in delta, gamma:** R. Masgrau, C. Guaza, R. M. Ransohoff, et al., "Should We Stop Saying 'Glia' and 'Neuroinflammation'?" *Trends in Molecular Medicine* 23, no. 6 (June 2017), 486–500.

116 **Delivered correctly, this noninvasive TMS:** C. L. Cullen and K. M. Young, "How Does Transcranial Magnetic Stimulation Influence Glial Cells in the Central Nervous System?" *Frontiers in Neural Circuits* 10 (April 2016), 26. According to Jay Gunkelman, a leader in qEEG research, direct stimulation in transcranial magnetic stimulation can help activate or deactivate circuitry and cause microglia to become active or inactive.

Nine: A Beleaguered Mind

120 **If the brain does not respond well early on:** M. S. Kelly, A. J. Oliveira-Maria, M. Bernstein, et al., "Initial Response to Transcranial Magnetic Stimulation Treatment for Depression Predicts Subsequent Response," *Journal of Neuropsychiatry and Clinical Neurosciences* 29, no. 2 (Spring 2017), 179–82.

120 **Dr. Asif has secured a respirator belt:** Here, and in other sections of this book where I describe Dr. Hasan Asif's work, I want to acknowledge that in addition to my own reporting, I also occasionally drew upon the reporting done by Amy Ellis Nutt, in her piece, "The Mind's Biology," *Washington Post*, February 19, 2016, A1–A14. You can read her piece here: www.washingtonpost.com/sf/national/2016/02/19/brain-hacking-the-minds-biology (accessed November 12, 2017).

122 **James Joyce's famous line:** James Joyce, *Dubliners* (New York: Penguin, 1993), 104.

Ten: Untangling Alzheimer's

139 **Scientists at the University of California, San Francisco, had already:** L. Verret, E. O. Mann, G. B. Hang, et al., "Inhibitory Interneuron Deficit Links Altered Network Activity and Cognitive Dysfunction in Alzheimer Model," *Cell* 149, no. 3 (April 27, 2012), 708–21; and A. K. Gillespie, E. A. Jones, Y.-H. Lin, et al., "Apolipoprotein E4 Causes Age-Dependent Disruption of Slow Gamma Oscillations during Hippocampal Sharp-Wave Ripples," *Neuron* 90, no. 4 (May 2016), 740–51. In 2018, researchers at the Karolinska Institute in Sweden also found that advancing cognitive decline is accompanied by progressive impairment of cognition-relevant EEG patterns in gamma oscilla-

tions. H. Balleza-Tapia, S. Crux, Y. Andrade-Talavera, et al., "TrpV1 Receptor Activation Rescues Neuronal Function and Network Gamma Oscillations from Aβ-Induced Impairment in Mouse Hippocampus in Vitro," *Elife* 7 (November 2018), e37703.

139 **Tsai had already found, in animal studies:** H. F. Iaccarino, A. C. Singer, A. J. Martorell, et al., "Gamma Frequency Entrainment Attenuates Amyloid Load and Modifies Microglia," *Nature* 540, no. 7632 (December 7, 2016), 230–35.

142 **Tsai published her findings first in the journal *Nature*:** Ibid. See also A. J. Martorell, A. L. Paulson, H.-J. Suk, et al., "Multi-Sensory Gamma Stimulation Ameliorates Alzheimer's-Associated Pathology and Improves Cognition," *Cell* (March 7, 2019) (epub ahead of print).

142 **Scientists at the Queensland Brain Institute:** G. Leinenga and J. Götz, "Scanning Ultrasound Removes Amyloid-β and Restores Memory in an Alzheimer's Disease Mouse Model," *Science Translational Medicine* 7, no. 278 (March 11, 2015), 278ra33.

143 **In 2018, Tsai published a paper sharing her team's instructions:** A. C. Singer, A. J. Martorell, J. M. Douglas, et al., "Noninvasive 40-Hz Light Flicker to Recruit Microglia and Reduce Amyloid Beta Load," *Nature Protocols* 13, no. 8 (August 2, 2018), 1850–68.

143 **Remember that in 2016, Margaret McCarthy's:** J. W. VanRyzin, S. J. Yu, M. Perez-Pouchoulen, et al., "Temporary Depletion of Microglia during the Early Postnatal Period Induces Lasting Sex-Dependent and Sex-Independent Effects on Behavior in Rats," *eNeuro* 3, no. 6 (November–December 2016), e0297–16. 2016, 1–19.

144 **Instead, this gene undergoes shifts:** S. Hong and B. Stevens, "TREM2: Keeping Microglia Fit during Good Times and Bad," *Cell Metabolism* 26, no. 4 (October 2017), 590–91; S. E. Hickman and J. El Khoury, "TREM2 and the Neuroimmunology of Alzheimer's Disease," *Biochemical Pharmacology* 88, no. 4 (April 15, 2014), 495–98; P. Yuan, C. Condello, C. D. Keene, et al., "TREM2 Haplodeficiency in Mice and Humans Impairs the Microglia Barrier Function Leading to Decreased Amyloid Compaction and Severe Axonal Dystrophy," *Neuron* 90, no. 4 (May 18, 2016), 724–39. For a longer discussion of specific variants in TREM2 that increase the risk of Alzheimer's, and the mechanisms by which this occurs, see Lisa Bain, Noam I. Keren, and Sheena M. Posey Norris, *Biomarkers of Neuroinflammation: Proceedings of a Workshop* (Washington, DC: National Academies Press, 2018), 17–19.

145 **Microglia were congregating and destroying:** B. Stevens, S. Hong, B. Barres, et al., "Complement and Microglia Mediate Early Synapse Loss in Alzheimer Mouse Models," *Science* 352, no. 6286 (May 6, 2016), 712–16; Bain, Keren, and Norris, *Biomarkers of Neuroinflammation*, 31.

145 **The correlation between synapses' disappearance:** Bain, Keren, and Norris, *Biomarkers of Neuroinflammation*, 19.

147 **Another way that scientists are interrogating:** E. Z. Macosko, A. Basu,

R. Satija, et al., "Highly Parallel Genome-Wide Expression Profiling of Individual Cells Using Nanoliter Droplets," *Cell* 161, no. 5 (May 21, 2015), 1202–14.

148 **Already, a team of researchers at Yale:** W. Bao, H. Jia, S. J. Finnema, et al., "PET Imaging for Early Detection of Alzheimer's Disease: From Pathologic to Physiologic Biomarkers," *PET Clinics* 12, no. 3 (July 2017), 329–50; S. J. Finnema, N. B. Nabulsi, J. Mercier, et al., "Kinetic Evaluation and Test-Retest Reproducibility of [^{11}C]UCB-J, a Novel Radioligand for Positron Emission Tomography Imaging of Synaptic Vesicle Glycoprotein 2A in Humans," *Journal of Cerebral Blood Flow and Metabolism* 38, no. 11 (November 2018), 2041–52.

150 **For instance, the researchers in Australia:** R. M. Nisbet, A. Van der Jeugd, G. Leinenga, et al., "Combined Effects of Scanning Ultrasound and a Tau-Specific Single Chain Antibody in a Tau-Transgenic Mouse Model," *Brain* 150, no. 5 (May 2017), 1220–30; G. Leinenga, C. Langton, R. Nisbet, et al., "Ultrasound Treatment of Neurological Diseases—Current and Emerging Applications," *Nature Reviews: Neurology* 12, no. 3 (March 2016), 161–74.

Eleven: Desperately Seeking Healthy Synapses

159 **Researchers at Johns Hopkins have shown:** M. Fotuhi, B. Lubinski, M. Trullinger, et al., "A Personalized 12-week 'Brain Fitness Program' for Improving Cognitive Function and Increasing the Volume of Hippocampus in Elderly with Mild Cognitive Impairment," *Journal of Prevention of Alzheimer's Disease* 3, no. 3 (2016), 133–37.

159 **Other studies have found, using fMRI scans:** J. Ghaziri, A. Tucholka, V. Larue, et al., "Neurofeedback Training Induces Changes in White and Gray Matter," *Clinical EEG and Neuroscience* 44, no. 4 (October 2013), 265–72.

159 **Similar studies show growth in cortical gray matter:** A. Munivenkatappa, J. Rajeswaram, B. Indira Devi, et al., "EEG Neurofeedback Therapy: Can It Attenuate Brain Changes in TBI?" *NeuroRehabilitation* 35, no. 3 (2014), 481–84.

159 **in one 2018 single-blind, randomized controlled trial:** D.M.A. Mehler, M. O. Sokunbi, I. Habes, et al., "Targeting the Affective Brain—A Randomized Controlled Trial of Real-Time fMRI Neurofeedback in Patients with Depression," *Neuropsychopharmacology* 43, no. 13 (December 2018), 2578–85.

159 **In a smaller pilot study, nearly half of patients with major depressive disorder:** F. Peeters, M. Oehlen, J. Ronner, et al., "Neurofeedback as a Treatment for Major Depressive Disorder—A Pilot Study," *PLOS One* 9, no. 3 (March 18, 2014), e91837; and R. Markiewcz, "The Use of EEG Biofeedback/Neurofeedback in Psychiatric Rehabilitation," *Psychiatria Polska* 51, no. 6 (December 30, 2017), 1095–106.

159 **In a 2016 randomized controlled trial, nearly three-quarters:** B. A. van der

Kolk, H. Hodgdon, M. Gapen, et al., "A Randomized Controlled Study of Neurofeedback for Chronic PTSD," *PLOS One* 11, no. 12 (December 2016), e0166752.

160 **Similar trials show a statistically:** S. Banerjee and C. Argáez, "Neurofeed-back and Biofeedback for Mood and Anxiety Disorders: A Review of Clinical Effectiveness and Guidelines," *CADTH Rapid Response Reports,* Canadian Agency for Drugs and Technologies in Health (November 2017). In one study, after fifteen sessions of neurofeedback, patients with generalized anxiety disorder reported better brain functioning and less anxiety, as compared to a control group. M. Dadashi, B. Birashk, F. Taremian, et al., "Effects of Increase in Amplitude of Occipital Alpha & Theta Brain Waves on Global Function-ing Level of Patients with GAD," *Basic and Clinical Neuroscience* 6, no. 1 (January 2015), 14–20.

160 **Other research has demonstrated that in a significant number of patients:** T. Sürmeli and A. Ertem, "Obsessive-Compulsive Disorder and the Efficacy of qEEG-Guided Neurofeedback Treatment: A Case Series," *Clinical EEG and Neuroscience* 42, no. 3 (July 2011), 195–201; F. Blaskovits, J. Tyerman, and M. Luctkar-Flude, "Effectiveness of Neurofeedback Therapy for Anxiety and Stress in Adults Living with a Chronic Illness: A Systematic Review Protocol," *JBI Database of Systematic Reviews and Implementation Reports* 15, no. 7 (July 2017), 1765–69; M. Luctkar-Flude and D. Groll, "A Systematic Review of the Safety and Effect of Neurofeedback on Fatigue and Cognition," *Integrative Cancer Therapies* 14, no. 4 (July 2015), 318–40; A. Munivenkatappa, J. Rajeswaran, N. Bennet, et al., "EEG Neurofeedback Therapy: Can It At-tenuate Brain Changes in TBI?" *NeuroRehabilitation* 35, no. 3 (2014), 481–84; and J. Ghaziri, A. Tucholka, V. Larue, et al., "Neurofeedback Training Induces Changes in White and Gray Matter," *Clinical EEG and Neuroscience* 44, no. 4 (October 2013), 265–72.

160 **Randomized controlled trials have shown that neurofeedback helped to alleviate:** J. Schmidt and A. Martin, "Neurofeedback Against Binge Eating: A Randomized Controlled Trial in a Female Subclinical Threshold Sample," *European Eating Disorders Review* 24, no. 5 (September 2016), 406–16; and J. Schmidt and A. Martin, "Neurofeedback Reduces Overeating Episodes in Female Restrained Eaters: A Randomized Controlled Pilot-Study," *Applied Psychophysiology and Biofeedback* 40, no. 4 (December 2015), 283–95.

160 **Several randomized controlled trials have shown improvements in symp-toms of ADHD:** F. Minder, A. Zuberer, D. Brandeis, et al., "Informant-Related Effects of Neurofeedback and Cognitive Training in Children with ADHD including a Waiting Control Phase: A Randomized-Controlled Trial," *European Child & Adolescent Psychiatry* 27, no. 8 (August 2018), 1055–66.

162 **For instance, patients who experience depression:** F. Peeters, M. Oehlen, J. Ronner, et al., "Neurofeedback as a Treatment for Major Depressive Disorder—A Pilot Study," *PLOS One* 9, no. 13 (March 2014), e91837; V. Zotev, R. Phillips, K. D. Young, et al., "Prefrontal Control of the Amyg-

dala During Real-Time fMRI Neurofeedback Training of Emotional Regulation," *PLOS One* 8, no. 11 (November 2013), e79184; and V. Zotev, H. Yuan, M. Misaki, et al., "Correlation between Amygdala BOLD Activity and Frontal EEG Asymmetry during Real-time fMRI Neurofeedback Training in Patients with Depression," *Neuroimage: Clinical* 11 (February 2016), 224–38.

162 **Patients with generalized anxiety disorder:** M. Dadashi, B. Birashk, F. Taremian, et al., "Effects of Increase in Amplitude of Occipital Alpha & Theta Brain Waves on Global Functioning Level of Patients with GAD," *Basic and Clinical Neuroscience* 6, no. 1 (January 2015), 14–20.

165 **"Using a noninvasive charge such as neurofeedback":** This is based on my phone interview with Jay Gunkelman on May 11, 2017.

166 **In 2015, researchers at the University of California, Irvine:** A. M. Taylor, A. W. Castonguay, A. J. Taylor, et al., "Microglia Disrupt Mesolimbic Reward Circuitry in Chronic Pain," *Journal of Neuroscience* 35, no. 22 (June 3, 2015), 8442–50; M. W. Salter and S. Beggs, "The Known Knowns of Microglia-Neuronal Signaling in Neuropathic Pain," *Neuroscience Letters* 557, pt. A (December 2013), 37–42.

166 **Also in 2015, researchers at Massachusetts General Hospital:** M. L. Loggia, D. B. Chonde, O. Akeju, et al., "Evidence for Brain Glial Activation in Chronic Pain Patients," *Brain* 138, pt. 3 (March 2015), 604–15.

166 **Neuroscientists have also found that when "accelerated":** A. M. Taylor, A. W. Castonguay, A. J. Taylor, et al., "Microglia Disrupt Mesolimbic Reward Circuitry in Chronic Pain," *Journal of Neuroscience* 35, no. 22 (June 3, 2015), 8442–50.

167 **One more interesting piece of research underscores how pain and emotional mood:** C. N. Dewall, G. Macdonald, G. D. Webster, et al., "Acetaminophen Reduces Social Pain: Behavioral and Neural Evidence," *Psychological Science* 21, no. 7 (July 2010), 931–37.

167 **This in turn sparks neuroinflammation, which contributes:** Z. Wu and H. Nakanishi, "Lessons from Microglia Aging for the Link between Inflammatory Bone Disorders and Alzheimer's Disease," *Journal of Immunology Research* 2015 (January), 471342.

167 **Once bone inflammation stokes brain inflammation:** F. R Nieto, A. K. Clark, J. Grist, et al., "Neuron-Immune Mechanisms Contribute to Pain in Early Stages of Arthritis," *Journal of Neuroinflammation* 13, no. 1 (April 29, 2016), 96; M. Fusco, S. D. Skaper, S. Coaccioli, et al., "Degenerative Joint Diseases and Neuroinflammation," *Pain Practice* 17, no. 4 (April 2017), 522–32.

169 **"These symptoms in the mind or body are indications":** This is based on my interview with Sebern Fisher, M.A., on April 14, 2018. You can read more in Fisher's book: Sebern F. Fisher, *Neurofeedback in the Treatment of Developmental Trauma: Calming the Fear-Driven Brain* (New York: W. W. Norton, 2014).

Twelve: Rebooting the Family Fixer

170 **How is it that a ballerina can execute:** BBC journalist Andrew Marr uses a similar ballerina analogy in his article "Marr's Mini Miracle," *Sunday Daily Mail*, January 21, 2017.

Thirteen: In Search of a Fire Extinguisher for the Brain

182 **First, he took a position as dean of research at Georgetown University:** At Georgetown, Faden also served as the founding director of Georgetown's Institute for Cognitive and Computational Sciences.

183 **Perhaps we've seen headlines reporting that autopsies performed on NFL:** J. Mez, P. T. Kiernan, B. Abolmohammadi, et al., "Clinicopathological Evaluation of Chronic Traumatic Encephalopathy in Players of American Football," *JAMA* 318, no. 4 (July 25, 2015), 360–70.

184 **Historically, the loss of cognitive function:** "The Old Man and the CTE: Did Brain Injury Lead to the Demise of Ernest Hemingway?" *Washington Post*, May 5, 2017, C2.

184 **Military vets—like Heather's husband, Dave:** H. Terrio, L. A. Brenner, B. J. Ivins, et al., "Traumatic Brain Injury Screening: Preliminary Finding in a US Army Brigade Combat Team," *Journal of Head Trauma Rehabilitation* 24, no. 1 (January–February 2009), 14–23.

184 **There are nearly four million head injuries:** www.npr.org/sections/health -shots/2016/05/31/479750268/poll-nearly-1-in-4-americans-report-having-had -a-concussion (accessed April 21, 2018).

185 **A single "moderate" traumatic brain injury:** G. Scott, A. F. Ramlack-hansingh, P. Edison, et al., "Amyloid Pathology and Axonal Injury After Brain Trauma," *Neurology* 86, no. 9 (March 1, 2016), 821–28.

185 **Children, girls, and women with brain injuries:** M. Albicini and A. McKinlay, "Anxiety Disorders in Adults with Childhood Traumatic Brain Injury: Evidence of Difficulties More than 10 Years Postinjury," *Journal of Head Trauma Rehabilitation* 33, no. 3 (May/June 2018), 191–99.

185 **The brains of those who've suffered concussions:** J. H. Cole, R. Leech, D. J. Sharp, et al., "Prediction of Brain Age Suggests Accelerated Atrophy After Traumatic Brain Injury," *Annals of Neurology* 77, no. 4 (April 2015), 571–81.

185 **In one recent study of 235,000 patients' health records:** M. Fralick, D. Thiruchelvam, H. C. Tien, et al., "Risk of Suicide After a Concussion," *Canadian Medical Association Journal* 188, no. 7 (April 19, 2016), 497–504.

185 **And, says Faden, more than 40 percent:** A. I. Faden and D. J. Loane, "Chronic Neurodegeneration After Traumatic Brain Injury: Alzheimer Disease, Chronic Traumatic Encephalopathy, or Persistent Neuroinflammation?" *Neurotherapeutics* 12, no. 1 (January 2015), 143–50.

186 **These mice also showed neurodegeneration:** D. J. Loane, A. Kumar, B. A.

Stoica, et al., "Progressive Neurodegeneration After Experimental Brain Trauma: Association with Chronic Microglial Activation," *Journal of Neuropathology & Experimental Neurology* 73, no. 1 (January 2014), 14–29.

186 **An injury to the head triggers microglia to switch:** A. I. Faden and D. J. Loane, "Chronic Neurodegeneration After Traumatic Brain Injury: Alzheimer Disease, Chronic Traumatic Encephalopathy, or Persistent Neuroinflammation?" *Neurotherapeutics* 12, no. 1 (January 2015), 143–50; C. K. Donat, G. Scott, S. M. Gentleman, et al., "Microglial Activation in Traumatic Brain Injury," *Frontiers in Aging Neuroscience* 9 (June 2017), 208.

187 **Astonishingly, some of these particles that microglia:** A. Kumar, B. A. Stoica, D. J. Loane, et al., "Microglial-Derived Microparticles Mediate Neuroinflammation After Traumatic Brain Injury," *Journal of Neuroinflammation* 14, no. 1 (March 15, 2017), 47.

188 **But spinal cord injuries, he says, can cause:** J. Wu, Z. Zhao, B. Sabirzhanov, et al., "Spinal Cord Injury Causes Brain Inflammation Associated with Cognitive and Affective Changes: Role of Cell Cycle Pathways," *Journal of Neuroscience* 34, no. 33 (August 13, 2014), 10989–11006.

188 **For instance, in one 2017 study, Swedish:** S. Montgomery, A. Hiyoshi, S. Burkill, et al., "Concussion in Adolescence and Risk of Multiple Sclerosis," *Annals of Neurology* 82, no. 4 (October 2017), 554–61.

190 **UCLA researchers recently found that exercise:** G. Krishna, R. Agrawal, Y. Zuang, et al., "7,8-Dihydroxyflavone Facilitates the Action Exercise to Restore Plasticity and Functionality: Implications for Early Brain Trauma Recovery," *Biochimica et Biophysica Acta* 1863, no. 6 (June 2017), 1204–13; A. Lal, S. A. Kolakowsky-Hayner, J. Ghajar, et al., "The Effect of Physical Exercise After a Concussion: A Systematic Review and Meta-Analysis," *American Journal of Sports Medicine* 46, no. 3 (March 2018), 743–52.

190 **Animal studies on intermittent fasting:** L. M. Davis, J. R. Pauly, R. D. Readnower, et al., "Fasting Is Neuroprotective Following Traumatic Brain Injury," *Journal of Neuroscience Research* 86, no. 8 (June 2008), 1812–22.

Fourteen: The Fast-er Cure?

195 **In fact, their brains appeared much younger:** Much of the information in this chapter comes from my interview with Valter Longo, as well as from reading the research papers cited below, and some of the information is derived from his book: Valter Longo, Ph.D., *The Longevity Diet* (New York: Avery, 2018).

196 **But more astonishingly, the diet promoted:** I. Y. Choi, L. Piccio, P. Childress, et al., "A Diet Mimicking Fasting Promotes Regeneration and Reduces Autoimmunity and Multiple Sclerosis Symptoms," *Cell Reports* 15, no. 10 (June 7, 2016), 2136–46.

196 **In animals on the fasting-mimicking diet:** I. Y. Choi, C. Lee, and V. D.

Longo, "Nutrition and Fasting Mimicking Diets in the Prevention and Treatment of Autoimmune Diseases and Immunosenescence," *Molecular and Cellular Endocrinology* 5, no. 455 (November 5, 2017), 4–12.

197 **Meanwhile, another leader in this growing niche:** V. D. Longo and M. P. Mattson, "Fasting: Molecular Mechanisms and Clinical Applications," *Cell Metabolism* 19, no. 2 (February 2014), 181–92.

197 **In one 2018 study, Mattson:** M. P. Mattson, K. Moehl, N. Ghena, et al., "Intermittent Metabolic Switching, Neuroplasticity, and Brain Health," *Nature Reviews: Neuroscience* 19, no. 2 (February 2018), 63–80.

198 **Over the past several years, more research on fasting:** J. B. Johnson, W. Summer, R. G. Cutler, et al., "Alternate Day Calorie Restriction Improves Clinical Findings and Reduces Markers of Oxidative Stress and Inflammation in Overweight Adults with Moderate Asthma," *Free Radical Biology and Medicine* 42, no. 5 (March 1, 2017), 665–74; M. P. Mattson, D. B. Allison, L. Fontana, et al., "Meal Frequency and Timing in Health and Disease," *Proceedings of the National Academy of Sciences* 111, no. 47 (November 25, 2014), 16647–53.

199 **In one stunning finding, Longo recently:** N. Guidi and V. D. Longo, "Periodic Fasting Starves Cisplatin-Resistant Cancers to Death," *EMBO Journal* 2018 (June) (epub ahead of print).

200 **Stacks of studies tell us that unhealthy alterations:** G. Winter, R. A. Hart, R.P.G. Charlesworth, et al., "Gut Microbiome and Depression: What We Know and What We Need to Know," *Reviews in the Neurosciences* (February 2018) (epub ahead of print).

200 **In one study of women between the ages of eighteen and fifty-six:** Z. Chen, J. Li, S. Gui, et al., "Comparative Metaproteomics Analysis Shows Altered Fecal Microbiota Signatures in Patients with Major Depressive Disorder," *NeuroReport* 29, no. 5 (March 2018), 417–25.

200 **The makeup of bacteria in the gut microbiome:** E. M. Glenny, E. C. Bulik-Sullivan, Q. Tang, et al., "Eating Disorders and the Intestinal Microbiota: Mechanisms of Energy Homeostasis and Behavioral Influence," *Current Psychiatry Reports* 19, no. 8 (August 2017), 51; T. R. Sampson, J. W. Debelius, T. Thron, et al., "Gut Microbiota Regulate Motor Deficits and Neuroinflammation in a Model of Parkinson's Disease," *Cell* 167, no. 6 (December 2016), 1469–80; E. Cekanaviciute, B. B. Yoo, T. F. Runia, et al., "Gut Bacteria from Multiple Sclerosis Patients Modulate Human T Cells and Exacerbate Symptoms in Mouse Models," *Proceedings of the National Academy of Sciences* 114, no. 40 (October 3, 2017), 10713–18.

200 **Chronic stress, which also promotes an inflammatory:** S. S. Yarandi, D. A. Peterson, G. J. Treisman, et al., "Modulatory Effects of Gut Microbiota on the Central Nervous System: How Gut Could Play a Role in Neuropsychiatric Health and Diseases," *Journal of Neurogastroenterology and Motility* 22, no. 2 (April 2016), 201–12.

200 **Our microbiome also directly affects levels of neurotransmitters:** N. W. Bellono, J. R. Bayner, D. B. F. Leitch, et al., "Enterochromaffin Cells Are Gut

Chemosensors That Couple to Sensory Neural Pathways," *Cell* 170, no. 1 (June 29, 2017), 185–98.

200 **Recently, scientists at the University of California:** In the description in the footnote on this page, I drew, in part, upon the reporting in the following article: "Researchers Learn More About How the Gut and Brain Interact," *Washington Post,* June 22, 2017, www.washingtonpost.com/news/to-your-health/wp/2017/06/22/our-gut-talks-and-sometimes-argues-with-our-brain-now-we-know-how (accessed June 19, 2018).

200 **This general understanding that microbes:** K. E. Sylvia and G. E. Demas, "A Gut Feeling: Microbiome-Brain-Immune Interactions Modulate Social and Affective Behaviors," *Hormones and Behavior* 99 (March 2018), 41–49.

201 **Patients with bipolar disorder:** F. Dickerson, M. Adamos, E. Katsafanas, et al., "Adjunctive Probiotic Microorganisms to Prevent Rehospitalization in Patients with Acute Mania: A Randomized Controlled Trial," *Bipolar Disorders* 20, no. 7 (November 2018), 614–21.

201 **When the gut microbiome is triggered:** I. Gabanyi, P. A. Muller, L. Feighery, et al., "Neuro-Immune Interactions Drive Tissue Programming in Intestinal Macrophages," *Cell* 164, no. 3 (January 28, 2016), 378–91.

201 **Microglia, in response, grow in size:** A. Castillo-Ruiz, M. Mosley, A. J. George, et al., "The Microbiota Influences Cell Death and Microglial Colonization in the Perinatal Mouse Brain," *Brain, Behavior, and Immunity* 67 (January 2018), 218–29.

208 **They discovered that intermittent fasting:** T. Okada, T. Otsubo, T. Hagiwara, et al., "Intermittent Fasting Prompted Recovery from Dextran Sulfate Sodium-Induced Colitis in Mice," *Journal of Clinical Biochemistry and Nutrition* 61, no. 2 (September 2017), 100–107.

208 **We know that patients with Crohn's disease:** K. Riazi, M. A. Galic, A. C. Kentner, et al., "Microglia-Dependent Alteration of Glutamatergic Synaptic Transmission and Plasticity in the Hippocampus During Peripheral Inflammation," *Journal of Neuroscience* 35, no. 12 (March 2015), 4942–52.

Fifteen: Future Medicine

211 **In one study, researchers analyzed data from 522 trials:** A. Cipriani, T. A. Furukawa, G. Salanti, et al., "Comparative Efficacy and Acceptability of 21 Antidepressant Drugs for the Acute Treatment of Adults with Major Depressive Disorder: A Systematic Review and Network Meta-Analysis," *Lancet* 391, no. 10128 (April 7, 2018), 1357–66.

211 **This data was conflicting given that an earlier:** I. Kirsch, "Antidepressants and the Placebo Effect," *Zeitschrift für Psychologie* 222, no. 3 (2014), 128–34.

211 **In an interesting aside, the so-called placebo effect:** You can read more about this fascinating new research on the placebo effect in Gary Greenberg's excellent article, "What If the Placebo Effect Isn't a Trick?" *New York*

Times Magazine, November 7, 2018, www.nytimes.com/2018/11/07/magazine/placebo-effect-medicine.html (accessed November 8, 2018).

211 **In two groups of patients with major depressive disorder:** V. H. Perry, "Microglia and Major Depression: Not Yet a Clear Picture," *Lancet Psychiatry* 5, no. 4 (April 2018), 292–94.

212 **Then a third study, which looked at depressed:** E. Setiawan, S. Attwells, A. A. Wilson, et al., "Association of Translocator Protein Total Distribution Volume with Duration of Untreated Major Depressive Disorder: A Cross-Sectional Study," *Lancet Psychiatry* 5, no. 4 (April 2018), 339–47.

212 **In one animal study, researchers found that after only a few weeks:** T. Kreisel, M. G. Frank, T. Licht, et al., "Dynamic Microglial Alterations Underlie Stress-Induced Depressive-Like Behavior and Suppressed Neurogenesis," *Molecular Psychiatry* 19, no. 6 (June 2014), 699–709.

212 **And some who do respond find that medications stop working for them over time:** World Health Organization, "Depression," www.who.int/en/news-room/fact-sheets/detail/depression (accessed September 4, 2018); National Institute of Mental Health, "Questions and Answers About the NIMH Sequenced Treatment Alternatives to Relieve Depression (STAR*D) Study—Background," January 2006, www.nimh.nih.gov/funding/clinical-research/practical/stard/backgroundstudy.shtml (accessed September 4, 2018).

212 **It may be that once microglia are engaged:** A. J. Rush, M. D. Madhukar, M. H. Trivedi, "Acute and Longer-Term Outcomes in Depressed Outpatients Requiring One or Several Treatment Steps: A STAR*D Report," *American Journal of Psychiatry* 163, no. 11 (November 2006), 1905–17.

213 **As the brain's ability to synthesize brain chemicals is diminished:** C. C. Watkins, A. Sawa, and M. G. Pomper, "Glia and Immune Cell Signaling in Bipolar Disorder: Insights from Neuropharmacology and Molecular Imaging to Clinical Application," *Translational Psychiatry* 4, no. 1 (January 2014), e350; G. Singhal and B. T. Baune, "Microglia: An Interface Between the Loss of Neuroplasticity and Depression," *Frontiers in Cellular Neuroscience* 11 (September 8, 2017), 270; and N. Herr, C. Bode, and D. Duerschmied, "The Effects of Serotonin in Immune Cells," *Frontiers in Cardiovascular Medicine* 4 (July 2017).

214 **But in other cases, too many microglia die off:** R. Yirmiya, N. Rimmerman, and R. Reshef, "Depression as a Microglial Disease," *Trends in Neuroscience* 38, no. 10 (October 2015), 637–58.

215 **In animal models, when healthy microglia are introduced back into the brain:** T. Kreisel, M. G. Frank, T. Licht, et al., "Dynamic Microglial Alterations Underlie Stress-Induced Depressive-Like Behavior and Suppressed Neurogenesis," *Molecular Psychiatry* 19, no. 6 (June 2014), 699–709.

215 **Since microglia have the ability to turn different gene functions on or off:** D. Gosselin, D. Skola, N. G. Coufal, et al., "An Environment-Dependent Transcriptional Network Specifies Human Microglia Identity," *Science* 356, no. 6344 (June 23, 2017), eaal3222.

215 **Many of the molecular pathways to depression:** T. Kreisel, M. G. Frank, R. Yirmiya, et al., "Dynamic Microglial Alterations Underlie Stress-Induced Depressive-Like Behavior and Suppressed Neurogenesism," *Molecular Psychiatry* 19, no. 6 (June 2014), 699–709.

215 **But what if we could find a way to safely stop these disease-associated genes:** P. Yaun, C. Condello, C. D. Keen, et al., "TREM2 Haplodeficiency in Mice and Humans Impairs the Microglia Barrier Function Leading to Decreased Amyloid Compaction and Severe Axonal Dystrophy," *Neuron* 90, no. 4 (May 18, 2016), 724–39.

216 **This gene predisposes the brain to mark too many synapses:** For more on this subject, see Amy Ellis Nutt's article "Scientists Open the 'Black Box' of Schizophrenia with Dramatic Genetic Discovery," *Washington Post*, January 27, 2016.

216 **Some labs are working to develop drugs for neuropsychiatric disorders:** Lisa Bain, Noam I. Keren, and Sheena M. Posey Norris, *Biomarkers of Neuroinflammation: Proceedings of a Workshop* (Washington, DC: National Academies Press, 2018), 33.

217 **And if there are ongoing inflammatory triggers:** P. K. Feltes, J. Doorduin, H. C. Klein, et al., "Anti-inflammatory Treatment for Major Depressive Disorder: Implications for Patients with an Elevated Immune Profile and Non-Responders to Standard Antidepressant Therapy," *Journal of Psychopharmacology* 31, no. 9 (September 2017), 1149–65.

217 **Researchers have shown that patients with higher levels of biomarkers:** E. Haroon, A. W. Daguanoo, B. J. Woolwine, et al., "Antidepressant Treatment Resistance Is Associated with Increased Inflammatory Markers in Patients with Major Depressive Disorder," *Psychoneuroendocrinology* 95 (September 2018), 43–49; D. R. Goldsmith, E. Haroon, A. H. Miller, et al., "Association of Baseline Inflammatory Markers and the Development of Negative Symptoms in Individuals at Clinical High Risk for Psychosis," *Brain, Behavior, and Immunity* 76 (February 2019), 269–74; M. Huang, S. Su, J. Goldberg, et al., "Longitudinal Association of Inflammation with Depressive Symptoms: A 7-Year Cross-Lagged Twin Difference Study," *Brain, Behavior, and Immunity* 75 (January 2019), 200–207; and A. H. Miller and C. L. Raison, "The Role of Inflammation in Depression: From Evolutionary Imperative to Modern Treatment Target," *Nature Reviews: Immunology* 16, no. 1 (January 2016), 22–34.

217 **So it makes good sense that combining anti-inflammatories with antidepressants:** N. Muller, M. J. Schwarz, S. Dehning, et al., "The Cyclooxygenase-2 Inhibitor Celecoxib Has Therapeutic Effects in Major Depression: Results of a Double-Blind, Randomized, Placebo-Controlled, Add-on Pilot Study to Reboxetine," *Molecular Psychiatry* 11, no. 7 (July 2006), 680–84.

217 **Taking the anti-inflammatory—which works to block the cytokine known as TNF:** Bain, Keren, and Norris, *Biomarkers of Neuroinflammation*, 34–35.

217 **Although the anti-TNF medication did not successfully:** Ibid.

217 **At Emory University, Andrew Miller, M.D.:** C. L. Raison, B. J. Woolwine,

R. E. Rutherford, et al., "A Randomized Controlled Trial of the Tumor Necrosis Factor Antagonist Infliximab for Treatment-Resistant Depression: The Role of Baseline Inflammatory Biomarkers," *JAMA Psychiatry* 70, no. 1 (January 2013), 31–41; A. H. Miller and C. L. Raison, "The Role of Inflammation in Depression: From Evolutionary Imperative to Modern Treatment Target," *Nature Reviews: Immunology* 16, no. 1 (January 2016), 22–34; J. C. Felger, E. Haroon, A. Patel, et al., "What Does Plasma CRP Tell Us About Peripheral and Central Inflammation in Depression?" *Molecular Psychiatry* (June 2018) (epub ahead of print); and A. H. Miller, M. Bekhbat, K. Chu, et al., "Glucose and Lipid-Related Biomarkers and the Antidepressant Response to Infliximab in Patients with Treatment-Resistant Depression," *Psychoneuroendocrinology* 98 (December 2018), 222–29.

218 **Miller is also working to bring greater precision:** A. H. Miller, M. H. Trivedi, and M. K. Jha, "Is C-Reactive Protein Ready for Prime Time in the Selection of Antidepressant Medications?" *Psychoneuroendocrinology* 84 (October 2017), 206.

218 **Another drug used for rheumatoid arthritis, tocilizumab:** B. J. Miller, J. K. Dias, H. P. Lemos, et al., "An Open-Label, Pilot Trial of Adjunctive Tocilizumab in Schizophrenia," *Journal of Clinical Psychiatry* 77, no. 2 (February 2016), 275–76.

218 **Studies are under way in the United Kingdom:** Hannah Devlin, "Radical New Approach to Schizophrenia Treatment Begins Trial," *The Guardian*, November 3, 2017, www.theguardian.com/society/2017/nov/03/radical-new-approach-to-schizophrenia-treatment-begins-trial (accessed August 20, 2018).

218 **It will take many more clinical trials before we clearly understand:** M. S. Cepeda, P. Stang, and R. Makadia, "Depression Is Associated with High Levels of C-Reactive Protein and Low Levels of Fractional Exhaled Nitric Oxide: Results from the 2007–2012 National Health and Nutrition Examination Surveys," *Journal of Clinical Psychiatry* 77, no. 12 (December 2016), 1666–71.

218 **165 patients who received monthly infusions of aducanumab for a year:** J. Sevigny, P. Chiao, T. Bussière, et al., "The Antibody Aducanumab Reduces Aβ Plaques in Alzheimer's Disease," *Nature* 537, no. 7618 (September 2016), 50–56.

219 **We know, for instance, that in fibromyalgia:** M. Ohgidani, T. A. Kato, M. Hosoi, et al., "Fibromyalgia and Microglia TNF-α: Translational Research Using Human Blood Induced Microglia-Like Cells," *Scientific Reports* 7, no. 1 (September 19, 2017), 11882.

219 **Among the common denominators in these disorders appear to be neuroinflammation:** Thanks to the blogger and science journalist Cort Johnson for his excellent reporting on ME/CFS and other rare disorders of the immune system.

220 **For instance, many patients with multiple sclerosis go on:** M. Srinivasan and D. K. Lahiri, "Significance of NF-κB as a Pivotal Therapeutic Target in

the Neurodegenerative Pathologies of Alzheimer's Disease and Multiple Sclerosis," *Expert Opinion on Therapeutic Targets* 19, no. 4 (April 2015), 471–87.

220 **Patients with fibromyalgia are twice as likely to develop dementia:** N.-S. Tzeng, C. H. Chung, F.-C. Liu, et al., "Fibromyalgia and Risk of Dementia—A Nationwide, Population-Based, Cohort Study," *American Journal of the Medical Sciences* 355, no. 2 (February 2018), 153–61.

220 **Young people who face high levels of anxiety and depression:** L. Mah, N. D. Anderson, N. P. L. G. Verhoeff, et al., "Negative Emotional Verbal Memory Biases in Mild Cognitive Impairment and Late-Onset Depression," *American Journal of Geriatric Psychiatry* 25, no. 10 (October 2017), 1160–70.

221 **The first glimmer came in the mid-1990s:** Gretchen Henkel, "Immune System No Longer Autonomous?" *The Rheumatologist*, May 1, 2010, www.the-rheumatologist.org/article/immune-system-no-longer-autonomous (accessed August 22, 2018).

222 **But after vagus nerve stimulation:** L. V. Borovikova, S. Ivanova, M. Zhang, et al., "Vagus Nerve Stimulation Attenuates the Systemic Inflammatory Response to Endotoxin," *Nature* 405, no. 6785 (May 2000), 458–62.

222 **Tracey's team recently found that low-level electrical vagal nerve stimulation:** F. A. Koopman, S. S. Chavan, S. Miljko, et al., "Vagus Nerve Stimulation Inhibits Cytokine Production and Attenuates Disease Severity in Rheumatoid Arthritis," *Proceedings of the National Academy of Sciences* 113, no. 29 (July 19, 2016), 8284–89. In this section detailing the work of Kevin Tracey, I also draw upon the reporting of Michael Behar in his informative article "Can the Nervous System Be Hacked?" *New York Times Magazine*, May 23, 2014.

223 **Vagal nerve stimulating devices have been used for some time:** L. Galbarriatu, I. Pomposo, A. Marinas, et al., "Vagus Nerve Stimulation Therapy for Treatment-Resistant Epilepsy: A 15-Year Experience at a Single Institution," *Clinical Neurology and Neurosurgery* 137 (October 2015), 89–93.

223 **And much of this research on vagal nerve stimulation:** J. P. Somann, G. O. Albors, K. V. Neihouser, et al., "Chronic Cuffing of Cervical Vagus Nerve Inhibits Efferent Fiber Integrity in Rat Model," *Journal of Neural Engineering* 15, no. 3 (June 2018), 036018.

223 **So they are experimenting with delivering low-level vagal nerve stimulation:** Y. A. Patel, T. Saxena, R. V. Bellamkonda, et al., "Kilohertz Frequency Nerve Block Enhances Anti-inflammatory Effects of Vagus Nerve Stimulation," *Scientific Reports* 7 (January 2017), 39810.

223 **A whole host of studies are now examining:** E. Meroni, N. Stakenborg, P. J. Gomez-Pinilla, et al., "Functional Characterization of Oxazolone-Induced Colitis and Survival Improvement by Vagus Nerve Stimulation," *PLOS One* 13, no. 5 (May 2018), e0197487; R. L. Johnson and C. G. Wilson, "A Review of Vagus Nerve Stimulation as a Therapeutic Intervention," *Journal of Inflammation Research* 11 (May 16, 2018), 203–13; G. S. Bassi, L. Ulloa, V. R.

Santos, "Cortical Stimulation in Conscious Rats Controls Joint Inflammation," *Progress in Neuro-Psychopharmacology & Biological Psychiatry* 84, pt. A, 201–13; K. Chakravarthy, H. Chaudhry, K. Williams, et al., "Review of the Uses of Vagal Nerve Stimulation in Chronic Pain Management," *Current Pain and Headache Reports* 19, no. 12 (December 2015), 54; C. Gaul, H. C. Diener, N. Silver, et al., "Non-invasive Vagus Nerve Stimulation for PREVention and Acute Treatment of Chronic Cluster Headache (PREVA): A randomized Controlled Study," *Cephalalgia* 36, no. 6 (May 2016), 534–46; H. I. Jacobs, J. M. Riphagen, C. M. Razat, et al., "Transcutaneous Vagus Nerve Stimulation Boosts Associative Memory in Older Individuals," *Neurobiology of Aging* 36, no. 5 (May 2015), 1860–67; A. Grimonprez, R. Raedt, J. Portelli, et al., "The Antidepressant-Like Effect of Vagus Nerve Stimulation Is Mediated Through the Locus Coeruleus," *Journal of Psychiatric Research* 68 (September 2015), 1–7; A. P. Shah, F. R. Carreno, H. Wu, et al., "Role of TrkB in the Anxiolytic-Like and Antidepressant-Like Effects of Vagal Nerve Stimulation," *Neuroscience* 322 (May 13, 2016), 273–86.

224 **Researchers in Denmark are also looking at how unhealthy changes:** E. Svensson, E. Horváth Puhó, R. W. Thomsen, et al., "Vagotomy and Subsequent Risk of Parkinson's Disease," *Annals of Neurology* 78, no. 4 (October 2015), 522–29.

224 **For instance, if the vagus nerve detects:** B. Bonaz, T. Bazin, and S. Pellissier, "The Vagus Nerve at the Interface of the Microbiota-Gut-Brain Axis," *Frontiers in Neuroscience* 12 (February 7, 2018), 49.

224 **He believes that when the vagal nerve itself is infected:** M. B. VanElzakker, "Chronic Fatigue Syndrome from Vagus Nerve Infection: A Psychoneuroimmunological Hypothesis," *Medical Hypotheses* 81, no. 3 (September 2013), 414–23.

224 **Increasingly, ketamine also appears to have the potential:** Q. Chen, J. Feng, L. Liu, et al., "The Effect of Ketamine on Microglia and Proinflammatory Cytokines in the Hippocampus of Depression-Like Rat," *Neuropsychiatry* 7, no. 2 (2017), 77–85.

225 **Although these microglia-specific findings have been in animal studies:** N. Diazgranados, L. Ibrahim, N. E. Brutsche, et al., "A Randomized Add-on Trial of an N-Methyl-D-Aspartate Antagonist in Treatment-Resistant Bipolar Depression," *Archives of General Psychiatry* 67, no. 8 (August 2010), 793–802; C. Rong, C. Park, J. D. Rosenblat, et al., "Predictors of Response to Ketamine in Treatment Resistant Major Depressive Disorder and Bipolar Disorder," *International Journal of Environmental Research and Public Health* 15, no. 4 (April 17, 2018), 771; A. K. Parsaik, B. Singh, D. Khosh-Chashm, et al., "Efficacy of Ketamine in Bipolar Depression: Systematic Review and Meta-Analysis," *Journal of Psychiatric Practice* 21, no. 6 (November 2015), 427–35; and S. J. Pennybaker, D. A. Luckenbaugh, C. A. Zarate, Jr., et al., "Ketamine and Psychosis History: Antidepressant Efficacy and Psychotomimetic Effects Postinfusion," *Biological Psychiatry* 82, no. 5 (September 2017), e35–e36.

225 **Susannah Tye, Ph.D., former director:** This is based on my Skype interview with Susannah Tye on December 18, 2018.

225 **Hospitals in the consortium:** For more information about the National Network of Depression Centers' Biomarker Discovery Project, see nndc.org/programs-events/biomarker-discovery-project/ (accessed January 5, 2019).

226 **Some physicians aren't waiting for all the clinical trials:** In this section I draw upon the reporting of Sarah Solovitch in her excellent article "Once-Popular Party Drug Now Used for Severe Depression," *Washington Post*, February 2, 2016, E1.

226 **The former director of the National Institute of Mental Health:** National Institute of Mental Health, "Post by Former NIMH Director Thomas Insel: Ketamine," October 1, 2014, www.nimh.nih.gov/about/directors/thomas-insel/blog/2014/ketamine.shtml (accessed September 10, 2018).

227 **That cost can diminish with treatment:** Ibid.

227 **Interestingly, patients who experience "emergence phenomena":** E. N. Aroke, S. L. Crawford, and J. R. Dungan, "Pharmacogenetics of Ketamine-Induced Emergence Phenomena: A Pilot Study," *Nursing Research* 66, no. 2 (March 2017), 105–14.

228 **In 2019, the FDA approved:** V. Popova, E. J. Daly, M. Trivedi, et al., "Randomized, Double-Blind Study of Flexibly-Dosed Intranasal Esketamine Plus Oral Antidepressant Versus Active Control in Treatment-Resistant Depression," presented at 2018 Annual Meeting of the American Psychiatric Association (APA), May 2018, New York, New York; E. J. Daly, M. Trivedi, A. Janik, et al., "A Randomized Withdrawal, Double-Blind, Multicenter Study of Esketamine Nasal Spray Plus an Oral Antidepressant for Relapse Prevention in Treatment-Resistant Depression," presented at the American Society of Clinical Psychopharmacology, May 2018, Miami, Florida. See also Benedict Carey, "Doctors Welcome New Depression Drug, Cautiously," *New York Times*, March 8, 2019.

228 **Another hallucinogen, psilocybin:** Robin Marantz Henig, "How a Psychedelic Drug Helps Cancer Patients Overcome Anxiety," December 3, 2016, NPR Shots. www.npr.org/sections/health-shots/2016/12/03/504136736/how-a-psychedelic-drug-helps-cancer-patients-overcome-anxiety (accessed September 6, 2018).

228 **Studies are under way to evaluate whether ayahuasca:** V. Dakic, R. M. Maciel, H. Drummond, et al., "Harmine Stimulates Proliferation of Human Neural Progenitors," *PeerJ* 4 (December 2016), e2727.

Sixteen: A Final Analysis

231 **Kipnis proposes that one of the defining roles:** Jonathan Kipnis, "The Seventh Sense," *Scientific American*, August 2018, 29–35.

231 **Researchers at the University of Colorado, Boulder:** M. D. Weber, M. G.

Frank, K. J. Tracey, et al., "Stress Induces the Danger-Associated Molecular Pattern HMGB-1 in the Hippocampus of Male Sprague Dawley Rats: A Priming Stimulus of Microglia and the NLRP3 Inflammasome," *Journal of Neuroscience* 35, no. 1 (January 2015), 316–24.

233 **Nearly a century ago, Sigmund Freud:** Helene Guldberg, "Review: *The Book of Woe*," *Psychology Today*, December 9, 2013.

235 **At one 2017 gathering of the top glial biologists:** Lisa Bain, Noam I. Keren, and Sheena M. Posey Norris, *Biomarkers of Neuroinflammation: Proceedings of a Workshop* (Washington, DC: National Academies Press, 2018), 12.

235 **Meanwhile, disorders of the mind and brain:** Michael McCarthy, "US Spent More on Mental Illness Than on Any Other Conditions in 2013, Study Finds," *BMJ* 353 (May 20, 2016), i2895.

236 **It has been said of individuals with psychiatric disorders:** Hope Jahren, *Lab Girl* (New York: Alfred A. Knopf, 2016), 49.

Index

About the Author

DONNA JACKSON NAKAZAWA is the author of three previous books exploring the intersection of neuroscience, immunology, and emotion: *Childhood Disrupted*, which was a finalist for the 2016 Books for a Better Life award, *The Last Best Cure*, and *The Autoimmune Epidemic*. For her written contributions to the field of immunity, she was the recipient of the 2012 AESKU award and the 2010 National Health Information Award, which recognizes the nation's best magazine articles in health. Jackson Nakazawa has appeared on *Today*, NPR, NBC, and ABC News, and her work has appeared in *The Washington Post*, *Health Affairs*, *Aeon*, *More*, *Parenting*, *AARP Magazine*, and *Glamour*, and has been highlighted on the cover of *Parade* as well as in *Time* and *USA Today*. She blogs for *Psychology Today* and *HuffPost*. Jackson Nakazawa has been the recipient of writing-in-residence fellowships from the Virginia Center for the Creative Arts, Yaddo, and the MacDowell Colony. She lives with her family in Maryland.

donnajacksonnakazawa.com
Facebook.com/donnajacksonnakazawaauthor
Twitter: @DonnaJackNak

About the Type

This book was set in Electra, a typeface designed for Linotype by W. A. Dwiggins, the renowned type designer (1880–1956). Electra is a fluid typeface, avoiding the contrasts of thick and thin strokes that are prevalent in most modern typefaces.